T0320665

Efficient Unconventional Models of Multi-Component Magnetic Hysteresis

Efficient Unconventional Models of Multi-Component Magnetic Hysteresis

Amr A Adly

Cairo University, Egypt

World Scientific

W JERSEY · LONDON · SINGAPORE · BEIJING · SHANGHAI · HONG KONG · TAIPEI · CHENNAI · TOKYO

Published by

World Scientific Publishing Co. Pte. Ltd.

5 Toh Tuck Link, Singapore 596224

USA office: 27 Warren Street, Suite 401-402, Hackensack, NJ 07601

UK office: 57 Shelton Street, Covent Garden, London WC2H 9HE

British Library Cataloguing-in-Publication Data
A catalogue record for this book is available from the British Library.

EFFICIENT UNCONVENTIONAL MODELS OF MULTI-COMPONENT MAGNETIC HYSTERESIS

ISBN 978-981-123-736-2 (hardcover)
ISBN 978-981-123-737-9 (ebook for institutions)
ISBN 978-981-123-738-6 (ebook for individuals)

For any available supplementary material, please visit
https://www.worldscientific.com/worldscibooks/10.1142/12291#t=suppl

Printed in Singapore

To the souls of my parents

Preface

It is well known that hysteresis is a phenomenon frequently encountered in many fields of specializations. Given that many physical quantities are in reality coupled, this book focuses on modeling multi-component hysteresis. While several rigorously mathematically formulated hysteresis models have been previously developed, a number of unconventional multi-component hysteresis approaches are presented in this book. General features of the proposed approaches are their computational efficiency, utilization of artificial neural networks, and the possibility to resolve their identification processes using any available set of experimental data. Overall, the book sheds the light on a collection of out-of-the-box modeling ideas that have evolved over a period of more than two decades. While introducing this book I should mention that I was fortunate to work on my PhD almost three decades ago under the supervision of Prof. Isaak Mayergoyz whose contributions related to the topic of hysteresis have been remarkable. I should also mention that without the interaction, collaboration, and brainstorming with many colleagues and fellow researchers the proposed ideas and methodologies would have never crystallized. Special thanks are due to Prof. Salwa Abd-El-Hafiz for her efforts in designing a long list of relevant neural networks. Finally, I hope that the content of this book would be useful to researchers working on relevant topics and I anticipate that the included ideas would stir more unconventional approaches to model multi-component hysteresis.

Amr A. Adly

Contents

List of Figures

Chapter 1

Introduction

Hysteresis is a phenomenon encountered in many fields of specializations. The most prominent examples where this phenomenon is encountered include magnetic hysteresis, superconducting hysteresis, chemical hysteresis and mechanical hysteresis (see, for instance, [1], [2], [3], [4] and [5]).

1.1. Scalar Hysteresis and the Primitive Hysteresis Operator

Scalar hysteresis may be best defined as a "multibranch nonlinearity for which branch-to-branch transitions occur after local extrema" [6]. Figure 1.1 clearly demonstrates this transition. In this figure, branching initiated from extremum Point X may trace different trajectories depending on the past local extrema history as depicted by Branch A and Branch B. Within the previous efforts to model this, more-or-less, complex phenomena a couple of basic properties associated with hysteresis were identified. Mainly, these two properties are the congruency and the wiping-out properties (see, for example, [7]). While this book mainly focuses on multi-component magnetic hysteresis, it is important to mention that the different phenomenological scalar hysteresis modeling strategies involved a weighted superposition of elementary hysteresis operators having different switching-up and switching-down thresholds as illustrated in Fig. 1.2. Samples of such operators are shown in Fig. 1.3 where α and β represent the switching thresholds. Examples of such models are presented in [8], [9], [6], [10], [11], [12], [13] and [14].

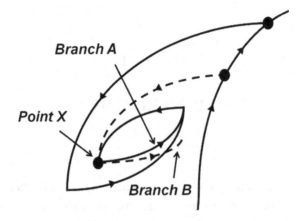

Figure 1.1. Demonstrating the history dependent branching in media exhibiting hysteresis.

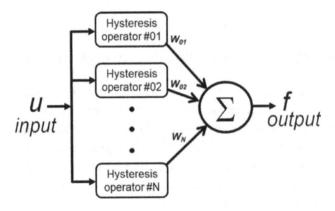

Figure 1.2. Construction of a scalar hysteresis model using an ensemble of elementary hysteresis operators.

In general, elementary hysteresis operators are characterized by their switching-up and switching-down thresholds designated by α and β as shown in Fig. 1.3. Obviously, a hysteresis model will include an ensemble of elementary operators having different switching up and down thresholds. While the simplest elementary hysteresis operator is the rectangular operator shown in Fig. 1.3.a, more sophisticated operators such as the Stoner-Wohlfarth-like operator given in Fig. 1.3.b have been also utilized (see, for instance, [13] and [15]). It should be stated here that as elementary hysteresis operators get more sophisticated, a smaller number of such operators would be usually needed to model hysteresis for a specific magnetic material.

For any hysteresis operator, a number of unknown parameters must be determined based upon some set of measured data that depends on the model structure. This action is usually referred to as the identification process. For an ensemble of pre-chosen set of elementary hysteresis operators, the identification process will reduce to the determination of the summing weights denoted by W_{ij} in Fig. 1.2. After the identification process is concluded, accuracy of the model may be assessed by its ability to predict input-output variations corresponding to measurements that were not included in the identification process.

It is worth mentioning here that the elementary hysteresis operator shown in Fig. 1.3.a may be electronically realized by a simple Schmitt trigger circuit (refer, for instance, to [16]). This fact is illustrated by a typical operational amplifier circuit shown in Fig. 1.4. Hence, by summing up weighted outputs of a group of Schmitt triggers whose switching up and down thresholds are appropriately tuned, a scalar hysteresis model may be physically realized. Such a physical realization may very well be utilized in real time compensation and/or control of systems exhibiting hysteresis.

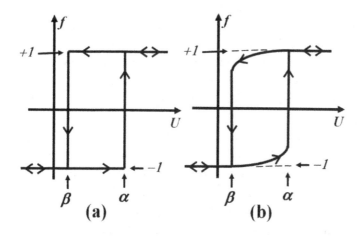

Figure 1.3. Examples of the elementary hysteresis operators.

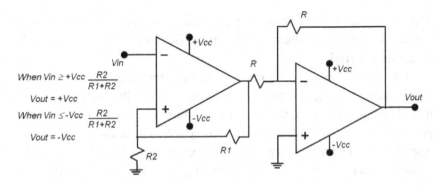

Figure 1.4. Realization of a symmetric elementary hysteresis operator using a typical operational amplifier Schmitt trigger circuit.

1.2. Multi-Component Magnetic Hysteresis

While most hysteresis models have been scalar in nature, fewer vector hysteresis models have been developed over the years. These vector models were either developed as inherently vector models such as the model reported in [5] or as some sort of generalization for pre-existing scalar models like those reported in [6], [17], [18] and [19]. In all cases, an invariant property of any vector hysteresis model is its ability to be utilized as a scalar hysteresis model if the applied field is restricted to vary along a single direction [6].

From a wider perspective, moving from scalar representation to a more general multi-component representation of magnetic hysteresis should not be reduced to a simple transition from scalar to vector magnetic hysteresis modeling. This is because the problem of magnetization assessment is indeed a coupled physical problem where the outcome is dependent on other non-magnetic factors. More specifically, assessment of magnetization as a result of some applied magnetic field should also take into consideration variation of other physical factors such the mechanical stress and the temperature (please refer to [20]). This fact was evident for those who used floppy disks few decades ago. In order to avoid data loss on a floppy disk due to misuse, typical instructions on its box included warnings against exposing the disk to stray fields, mechanical stresses resulting from bending the disk and excessive heat (see Fig. 1.5).

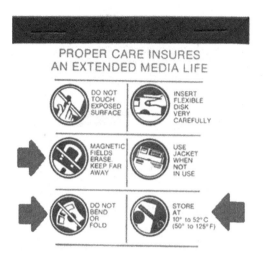

Figure 1.5. Typical care instructions listed on a floppy disk encloser (Athana Brand).

In block diagram terminology, input-output relations involving magnetic hysteresis may be generally represented as shown in Fig. 1.6. In accordance with the previous physical reasoning, inputs include magnetic field component(s), temperature and mechanical stress. On the other hand, outputs include magnetization component(s) and mechanical strain. Please note that temperature could also be included as an output component in the case when the magnetocaloric effect is considered (see for example [21], [22], [23], [24], [25] and [26]). Nevertheless, this effect has not been taken into consideration since it is mainly pronounced for a limited set of magnetic materials. Moreover, vector modeling of mechanical stress-strain components has also been ignored due to its limited field of applicability.

Based upon the basic physical facts related to the magnetization process [20] and referring to the block diagram shown in Fig. 1.6, it might be useful to mention the following points:

In the case when modeling is solely confined to scalar or vector magnetic hysteresis as a function of applied magnetic field, it should be clearly understood that either temperature and applied mechanical stress values are assumed to be maintained unchanged or that this modeling ignores their effects as an approximation.

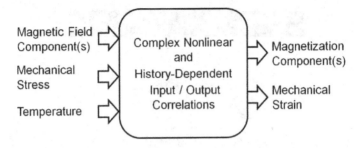

Figure 1.6. A high-level multi-component magnetic hysteresis input-output block diagram.

In the case when modeling is solely focused on thermal effects on magnetic field versus magnetization variations, it should be clearly understood that the mechanical stress is assumed to be maintained unchanged or that this modeling involves ignoring mechanical effects as an approximation.

In the case when modeling is solely focused on magnetostriction (i.e., magnetic field and applied mechanical stress versus mechanical strain), it should be clearly understood that the temperature is assumed to be maintained unchanged or that this modeling involves ignores temperature effects as an approximation.

Moreover, even though magnetic field, temperature and mechanical stress represent different physical quantities, they could very well have similar effects on magnetization. Generally speaking, variation of any of the aforementioned quantities could result in an irreversible change in the magnetization value. Furthermore, residual magnetization along a specific direction may be lost as a result of any of the following (Fig. 1.7):

- A sufficiently large magnetic field along the orthogonal direction.
- Temperature increase up to the Curie temperature limit [20].
- A sufficiently large mechanical stress along the same direction.

It should be pointed out here that Fig. 1.6 represents a clear justification for the typical care instructions associated with data preservation on magnetic recording media as previously demonstrated in Fig. 1.5.

Interdependence between magnetic and mechanical quantities is once more evident in materials exhibiting pronounced magnetostriction. For such materials, dimensions parallel to the applied magnetic field exhibit some variation which is proportional to the absolute value of that field. This variation (i.e., mechanical strain) could be positive or negative depending on whether the material is classified among those exhibiting positive or negative magnetostriction, respectively (refer, for instance, to [27]). In plain words, and as suggested by Fig. 1.6, mechanical strain may be changed by varying the mechanical stress, the magnetic field or both. For the case when the mechanical strain change is solely driven by cyclic magnetic field variation, a typical butterfly-type relation is always expected as shown in Fig. 1.8.

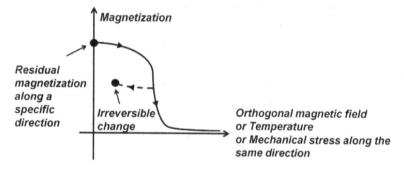

Figure 1.7. Qualitative decay of residual magnetization as a result of temperature increase or applied orthogonal field or applied mechanical stress.

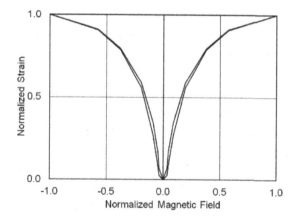

Figure 1.8. Typical butterfly-type curve correlating mechanical strain to a cyclic magnetic field variation for a magnetostrictive material.

1.3. Identification and Testing of Multi-Component Hysteresis

In general, any hysteresis model would include a set of unknown parameters that have to be determined, an activity referred to as the identification process. Since the model should be, in general, capable of simulating the input-output interactions under consideration for different materials, the identification process should be carried out for the specific material under consideration. Referring to the most widely used and mathematically well formulated hysteresis models, it can be realized that the identification process involves matching specific model unknowns using pre-identified set of measurements (refer, for instance, to [6], [28], [29], [30] and [31]). Unfortunately, some of the above-mentioned measurements require somewhat sophisticated experimental setups that might not be usually available within many research and/or industrial facilities. The question becomes whether it is possible to identify well established hysteresis models without the need to acquire, or at least access, the required experimental setups. Since this is not possible using conventional modeling and/or identification approaches, the need to follow unconventional means becomes quite clear.

1.4. Unconventional Multi-Component Hysteresis Modeling

Following up on the reasoning of the previous section, it becomes evident that unconventional hysteresis modeling approaches could be extremely useful. More specifically, main goals of exploring unconventional hysteresis approaches should basically include:

- Identification of widely used hysteresis models using any available set of measured or reported experimental data that might not match exactly the models pre-defined set of measurements.
- Introducing unconventional blocks that may partially or totally substitute some of the well-known hysteresis model building blocks.
- Achieving more computationally efficient multi-component hysteresis models that may be favorable in case such models are incorporated in field computation modules involving massive geometrical configuration discretization.

- Opening possibilities of utilizing widely available mathematical simulation modules in the construction of multi-component hysteresis models.

In the following chapters of this book different unconventional multi-component phenomenological hysteresis modeling approaches, covering problems of different physical natures, are discussed. Many of these approaches will partially involve unconventional methodologies to construct the elementary hysteresis operator. Since the afore-mentioned approaches are applicable to the various physical applications under consideration, a couple of innovative methodologies to construct elementary hysteresis operators are presented in the following Sections.

1.5. Realization of the Elementary Rectangular Hysteresis Operator Using Dual-Node Discrete Hopfield Neural Network

Let us consider the dual-node Hopfield neural network (HNN) shown in Fig. 1.9 and having discrete activation functions with binary outputs $\in \{-1, +1\}$ (see, for example, [32] and [33]). Given the nature of the activation functions, such a network falls within the discrete Hopfield neural network (DHNN) classification. As can be seen from Fig. 1.9, the external DHNN input is denoted by u, while the outputs are denoted by f_A and f_B for nodes A and B, respectively. Please note that for the

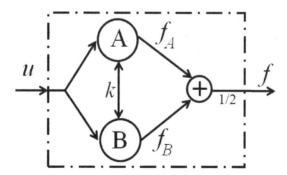

Figure 1.9. A dual node Hopfield neural network having positive feedback.

proposed implementation, positive feedback k between the DHNN nodes is assumed.

In accordance with the DHNN intrinsic mechanism, each node applies its step function to the sum of the external input u and the weighted output of the other node, resulting in an output $\in \{-1, +1\}$. Node outputs undergo changes, dependent on the external input u, until the DHNN state converges to a minimum of the energy function E given by [32]:

$$E = -\left[u(f_A + f_B) + k f_A f_B\right]. \tag{1.1}$$

Please note that according to the gradient descent rule the output states of nodes A and B are changed – for the same input instant – as per the following expressions:

$$f_A(t+1) = \text{sgn}(k f_B(t) + u) \tag{1.2}$$

$$f_B(t+1) = \text{sgn}(k f_A(t) + u) \tag{1.3}$$

where " sgn " represents the signum function, defined by:

$$\text{sgn}(x) = \begin{cases} +1 & \text{if } x > 0 \\ -1 & \text{if } x < 0 \\ \text{Unchanged} & \text{if } x = 0 \end{cases} \tag{1.4}$$

It turns out that the evolution track for outputs f_A and f_B in addition to averaging both outputs will yield the rectangular loop shown in Fig. 1.10. Note that the width of this loop is equivalent to twice the value of the DHNN positive feedback k. In case an offset component u_{os} is superimposed to the input u, the switching thresholds of the rectangular loop may be correlated to those of the loop given in Fig. 1.3 as follows:

$$(\alpha - \beta) = 2k \tag{1.5}$$

$$(\alpha + \beta)/2 = u_{os} \tag{1.6}$$

$$\alpha = u_{os} + k, \quad \beta = u_{os} - k \tag{1.7}$$

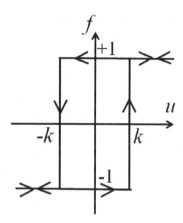

Figure 1.10. Elementary rectangular hysteresis operator generated by the dual-node DHNN shown in Fig. 1.9.

1.6. Realization of a Stoner-Wohlfarth like Hysteresis Operator Using Dual-Node Hopfield Neural Network Having a Hybrid Discrete-Continuous Activation Function

Although classified as a single domain micromagnetic model, the Stoner-Wohlfarth hysteresis model may be regarded as more general primitive operator [5]. A typical hysteresis curve generated in accordance with Stoner-Wohlfarth model is shown in Fig. 1.11. It should be stated that more general hysteresis models were constructed using an ensemble of Stoner-Wohlfarth operators having different parameters following

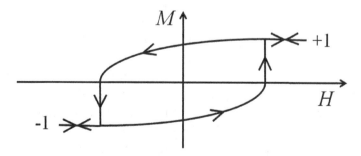

Figure 1.11. A typical hysteresis curve generated in accordance with the Stoner-Wohlfarth model.

configurations similar to that shown in Fig. 1.2 (refer, for instance, to [13]). Please note that the smooth monotonic increase and decrease in certain portions of this curve suggests that a smaller number of primitive operators of this type would be needed in comparison with rectangular operators to construct the same general hysteresis model.

According to the Stoner-Wohlfarth model for a single domain particle having a uniaxial anisotropy along the x-axis direction, the orientation φ of the magnetic moment M may be determined through the determination of the minimum value of the following energy E equation:

$$E = K \sin^2 \varphi - MH \cos \varphi \qquad (1.8)$$

where, K is the anisotropy constant while H is a unidirectional applied field along the x-axis.

It turns out that a primitive Stoner-Wohlfarth-like hysteresis operator offering more flexibility may be constructed using a dual node Hopfield neural network like that shown in Fig. 1.9 (refer, for instance, to [15] and [34]).

Consider an activation function f comprised of a weighted super-position of; a discrete activation function fd and a continuous one fc as given below:

$$f(x) = c\, fc(x) + (1-c)\, fd(x), \quad 0 \le c \le 1 \qquad (1.9)$$

$$fd(x) = \mathrm{sgn}(x) = \begin{cases} +1 & \text{if } x > 0 \\ -1 & \text{if } x < 0 \\ \text{Unchanged} & \text{if } x = 0 \end{cases} \qquad (1.10)$$

$$fc(x) = \tanh(ax), \quad 0 < a \qquad (1.11)$$

For a given input u, each node constantly updates its output according to the previous expressions. Due to the positive feedback between the nodes, output values may change until the network converges to a minimum value for its energy function given by equation (1.1).

Following the gradient descent rule for the discrete activation function component, the output of say node A is changed as:

$$f_A(t+1) = fd(x(t)) = \text{sgn}(k f_B(t) + u) \qquad (1.12)$$

Using the same gradient descent rule for the continuous activation function component, the output is changed gradually as:

$$\frac{\delta f_A}{\delta t} = \eta \, fc(x(t)) = \eta \tanh\big(a(k f_B(t) + u)\big) \qquad (1.13)$$

where, η is a small positive learning rate that controls the convergence speed (refer, for instance, [32]).

Sample Stoner-Wohlfarth-like hysteresis operators generated using the previously discussed HNN methodology are shown in Fig. 1.12 for different values of k and c. The figure clearly illustrates how the proposed hybrid activation function for the dual-node HNN results in loops with controllable width and squareness.

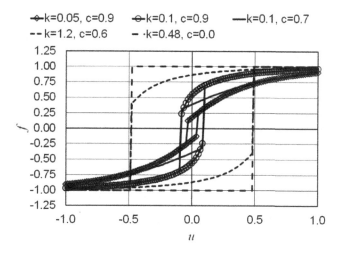

Figure 1.12. Sample Stoner-Wohlfarth-like hysteresis operators generated using the previously discussed HNN methodology for $a = 1$.

1.7. Prelude to The Following Chapters

In the following chapters of this book different unconventional multi-component phenomenological hysteresis modeling approaches, covering problems of different physical natures, are presented. More specifically, Chapter Two is devoted to the modeling of vector field variations, while Chapters Three and Four focus on the modeling of field-temperature and field mechanical variations, respectively. Finally, some conclusions are listed in Chapter Five. Given the different physical nature of the topics covered in these chapters, discussion within every chapter will tend to be self-contained. In other words, different approaches, their building blocks as well as identification strategies will be presented in detail for the physical problem under consideration irrespective of analogies that might exist in other chapters of this book.

Chapter 2

Magnetic Hysteresis Models for Vector Field Variations

Vector magnetic hysteresis models deal with multi-directional field-magnetization variations. The goal is always to end up with models which are capable qualitatively and quantitatively to mimic the previously mentioned multi-directional interactions. Obviously, accuracy extent, the ease of identification and computational efficiency represent comparative criteria for the selection of the most suitable model given a specific application. While vector hysteresis models may cover two-dimensional and three-dimensional interactions, discussions will be confined to the most encountered two-dimensional case. It should be pointed out here that extension of the presented modeling reasoning and methodologies may be easily inferred.

As previously stated in [6], vector magnetic hysteresis may be defined as "a vector nonlinearity with the property that past extrema of input projections along all possible directions may affect future output value". As previously stated in Chapter One, unconventional hysteresis models presented in this book are confined to phenomenological ones. In this sequel, it might be appropriate to list the main experimental facts associated with vector magnetic hysteresis (refer, for instance, to [6], [17], [18] and [28]). These facts may be summarized as follows:

- Reduction of the vector model field-magnetization variations to a scalar model when the magnetic field is restricted to vary along a single direction. In such a case, both the wiping-out and congruency properties should be satisfied [7]. It should be stated here that the unidirectional field-magnetization variations should be orientation-independent for isotropic media. Alternatively, these variations will be orientation-dependent for anisotropic media.

- Rotational symmetry property where a rotational magnetic field having a fixed amplitude would result in a superposition of fixed and rotational magnetization components whose relative magnitudes are dependent on the input rotational amplitude (see Fig. 2.1). The rotational magnetization component, in this case, tends to trace either a circular or an elliptical trajectory for isotropic and anisotropic media, respectively.

- Property of correlation between mutually orthogonal field-magnetization components. Note that this property has been briefly mentioned in Section 1.2 (see Fig. 1.7). According to this property, the magnetization along a specific direction may be altered, or even eliminated, due of the application of a magnetic field component along a different direction. As a special case, elimination of residual magnetization along a certain direction may be achieved by applying sufficiently large orthogonal magnetic field. Obviously, this is quite in line with the micromagnetic concept of domain rotation (refer, for instance, to [20]).

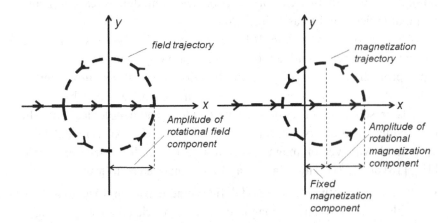

Figure 2.1. Magnetization components of an isotropic material resulting from the application of a rotational field component.

Obviously, widely used previously developed vector hysteresis models qualitatively satisfy the above main experimental facts. Accuracy of any of those models, on the other hand, is mainly judged by the quantitative agreement with experimental results. In Section 1.4, the different motivations to explore unconventional methodologies to model multi-component hysteresis have been stated. In the remaining part of this Chapter, several different unconventional approaches relevant to the modeling of vector hysteresis and/or identification of previously developed models are presented. Main features, mathematical implementation, identification techniques and experimental testing are also given.

2.1. Utilizing Feed-Forward Neural Networks in The Representation and Identification of Vector Preisach Models

It has been mentioned in the previous chapter that the identification of many widely used hysteresis models requires pre-defined set of measurements that might not necessarily be available among many potential users of these models. This fact has led to the investigation of the possibility of utilizing artificial neural networks (ANN) in the inference of model unknowns from the available experimental data (see, for instance, [35], [36], [37], [38] and [39]). This section presents an approach through which ANN may be utilized in the representation of a general anisotropic vector Preisach-type model form (in accordance with [18] and [40]) and its identification using any available measured vector data.

A general formulation for the 2D field-magnetization vector Preisach-type models may be given by:

$$M_x(t) = \iint_{\alpha \geq \beta} \left\{ \int_{-\pi/2}^{+\pi/2} \cos\varphi \; \zeta_x(\alpha,\beta) g_x(\varphi) \hat{\gamma}_{\alpha\beta} \left[\bar{e}_\varphi \bullet \bar{H}(t) \right] d\varphi \right\} d\alpha \; d\beta$$

$$(2.1)$$

$$M_y(t) = \iint_{\alpha \geq \beta} \left\{ \int_{-\pi/2}^{+\pi/2} \sin\varphi \; \zeta_y(\alpha,\beta) g_y(\varphi) \hat{\gamma}_{\alpha\beta} \left[\bar{e}_\varphi \bullet \bar{H}(t) \right] d\varphi \right\} d\alpha \; d\beta$$

$$(2.2)$$

where, t is the input-output time instant, $\hat{\gamma}_{\alpha\beta}$ represents the elementary hysteresis operator which when applied to an input yields the rectangular loop shown in Fig. 1.3.a, $\bar{e}_{\varphi}$ is a unit vector having an orientation identified by the direction polar angle φ, while M_x and M_y are the magnetization components along the x- and y-directions, respectively. In $(2.1) - (2.2)$, the functions ζ_x, ζ_y and the even functions g_x, g_y are the model unknowns required to be determined via the identification process.

By using first-order Fourier expansion approximation, the unknown even functions may be expressed in the forms:

$$g_x(\varphi) = g_{x0} + g_{x1}\cos\varphi \tag{2.3}$$

$$g_y(\varphi) = g_{y0} + g_{y1}\cos\varphi \tag{2.4}$$

Substituting $(2.3) - (2.4)$ into $(2.1) - (2.2)$ we get:

$$M_x(t) = \iint_{\alpha \geq \beta} \left\{ \int_{-\pi/2}^{+\pi/2} \cos\varphi \left(\begin{matrix} \zeta_{x0}(\alpha,\beta) \\ +\zeta_{x1}(\alpha,\beta)\cos\varphi \end{matrix} \right) \hat{\gamma}_{\alpha\beta}\left[\bar{e}_{\varphi} \bullet \bar{H}(t)\right] d\varphi \right\} d\alpha\, d\beta$$

$$\tag{2.5}$$

$$M_y(t) = \iint_{\alpha \geq \beta} \left\{ \int_{-\pi/2}^{+\pi/2} \sin\varphi \left(\begin{matrix} \zeta_{y0}(\alpha,\beta) \\ +\zeta_{y1}(\alpha,\beta)\cos\varphi \end{matrix} \right) \hat{\gamma}_{\alpha\beta}\left[\bar{e}_{\varphi} \bullet \bar{H}(t)\right] d\varphi \right\} d\alpha\, d\beta$$

$$\tag{2.6}$$

where,

$$\zeta_{x0}(\alpha,\beta) = g_{x0}\ \zeta_x(\alpha,\beta)\,,\quad \zeta_{x1}(\alpha,\beta) = g_{x1}\ \zeta_x(\alpha,\beta) \tag{2.7}$$

$$\zeta_{y0}(\alpha,\beta) = g_{y0}\ \zeta_y(\alpha,\beta)\,,\quad \zeta_{y1}(\alpha,\beta) = g_{y1}\ \zeta_y(\alpha,\beta) \tag{2.8}$$

Obviously, only a finite set of hysteresis operators and N angular orientations may be considered in the numerical implementation of the model. Denoting the maximum field under consideration by H_{sat}, (2.5) and (2.6) may be written in the form:

$$M_x(t) \approx \sum_{\alpha_i \geq \beta_j} \zeta_{x0}(\alpha_i,\beta_j) Sx_{\alpha_i\beta_j}^{(0)} + \sum_{\alpha_i \geq \beta_j} \zeta_{x1}(\alpha_i,\beta_j) Sx_{\alpha_i\beta_j}^{(1)} \tag{2.9}$$

$$M_y(t) \approx \sum_{\alpha_i \geq \beta_j} \zeta_{y0}(\alpha_i, \beta_j) Sy_{\alpha_i \beta_j}^{(0)} + \sum_{\alpha_i \geq \beta_j} \zeta_{y1}(\alpha_i, \beta_j) Sy_{\alpha_i \beta_j}^{(1)} \quad (2.10)$$

where,

$$1 \leq i, j \leq P, \ P = \text{odd number} \quad (2.11)$$

$$Sx_{\alpha_i \beta_j}^{(0)} = \left\{ \sum_{n=1}^{N} \cos \varphi_n \ \hat{\gamma}_{\alpha_i \beta_j} \left[\overline{e}_{\varphi_n} \bullet \overline{H}(t) \right] \frac{\pi}{N} \right\} \Delta \alpha \Delta \beta \quad (2.12)$$

$$Sx_{\alpha_i \beta_j}^{(1)} = \left\{ \sum_{n=1}^{N} \cos^2 \varphi_n \ \hat{\gamma}_{\alpha_i \beta_j} \left[\overline{e}_{\varphi_n} \bullet \overline{H}(t) \right] \frac{\pi}{N} \right\} \Delta \alpha \Delta \beta \quad (2.13)$$

$$Sy_{\alpha_i \beta_j}^{(0)} = \left\{ \sum_{n=1}^{N} \sin \varphi_n \ \hat{\gamma}_{\alpha_i \beta_j} \left[\overline{e}_{\varphi_n} \bullet \overline{H}(t) \right] \frac{\pi}{N} \right\} \Delta \alpha \Delta \beta \quad (2.14)$$

$$Sy_{\alpha_i \beta_j}^{(1)} = \left\{ \sum_{n=1}^{N} \frac{\sin 2\varphi_n}{2} \ \hat{\gamma}_{\alpha_i \beta_j} \left[\overline{e}_{\varphi_n} \bullet \overline{H}(t) \right] \frac{\pi}{N} \right\} \Delta \alpha \Delta \beta \quad (2.15)$$

$$\varphi_n = -\frac{\pi}{2} + \left(n - \frac{1}{2} \right) \frac{\pi}{N} \quad (2.16)$$

The identification problem, thus, reduces to the determination of the unknowns ζ_{x0}, ζ_{x1}, ζ_{y0} and ζ_{y1} as observed from equations (2.9) and (2.10). It can be easily realized that the two feed-forward artificial neural networks (FFANN) shown in Fig. 2.2 may separately represent equations (2.9) and (2.10). Each of these FFANN is comprised of an input stage, a single node hidden layer and a single node output layer. According to the aforementioned equations, the number of input stage is twice the number of utilized elementary hysteresis (i.e., twice $(P(P+1)/2)$). For the hidden and output layer nodes, the following piecewise linear node function $f(x)$ is employed:

$$f(x) = \max(-f_{sat}, \min(x, +f_{sat})), \ f_{sat} > \max(M_x, M_y) \quad (2.17)$$

Adopting this linear activation function selection offers the ability to explicitly extract the model unknowns from the various FFANN branch weights after the training process is terminated successfully.

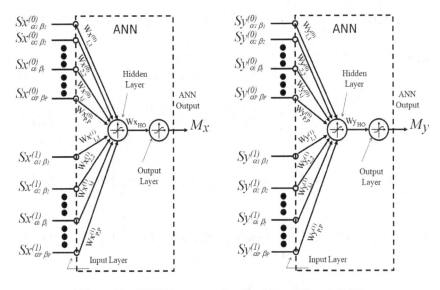

Figure 2.2.   FFANNs representing Equations (2.9) and (2.10).

More explicitly, identification of the model may be carried out through the training processes of the FFANNs using the available normalized measured vector data. Obviously, accuracy of the identification process is dependent on having available any arbitrary measured data that, whether implicitly or explicitly, reflect the magnetic properties of the material under consideration. First, summations $Sx_{\alpha_i\beta_j}^{(0)}$, $Sx_{\alpha_i\beta_j}^{(1)}$, $Sy_{\alpha_i\beta_j}^{(0)}$ and $Sy_{\alpha_i\beta_j}^{(1)}$ are performed to transform the actual field sequence into a corresponding sequence of ANNs input vectors x_q.

During the training process these input vectors are passed to the hidden layer node of each FFANN which, consequently, computes a weighted sum of its inputs and passes the sum through its activation function to the output layer. Likewise, the output node computes the network output. The initially randomly set weights $Wx^{(0)}_{i,j}$, $Wy^{(0)}_{i,j}$, $Wx^{(1)}_{i,j}$, $Wy^{(1)}_{i,j}$, Wx_{HO} and Wy_{HO} of the ANN branches are continuously modified during the training process until the closest possible match between the ANN output sequence o_q and the measured output sequence d_q for all the input vector sequences x_q. This training employs a supervised back propagation learning algorithm which targets the minimization of the quadratic error $Err = \dfrac{1}{2}\sum_q \left(o_q - d_q\right)^2$ that is dependent on the various network weights.

The gradient descent method is used for error minimization and the modification of network weights is carried out in a direction corresponding to the error *Err* negative gradient. Error at the output layer is propagated backwards to the hidden layer by using the weighted error transferred to the hidden node from the output node and the desired weight modifications for individual connections ending at the hidden node are calculated. Upon successful completion of the training phase for both FFANNs, the network weights are frozen. At this stage, the FFANNs may be used to predict M_x and M_y corresponding to arbitrary $\bar{H}$ variations. In case the FFANNs role is only confined to the identification of the vector Preisach mode unknowns rather than making use of them in the simulation phase, the model unknowns may, accordingly, be deduced from the expressions:

$$\zeta_{uv}(\alpha_i,\beta_j) = Wu_{HO} \times Wu_{i,j}^{(v)}, \text{ where } u = x, y \text{ and } v = 0, 1 \quad (2.18)$$

It should be mentioned here that the classical isotropic vector Preisach-type model, corresponds to the special case when $\zeta_x = \zeta_y$ and $g_x(\varphi) = g_y(\varphi) = 1$. In this situation, the model is usually identified using scalar measured data only and, hence, the FFANN shown in Fig. 2.3 would be needed to find the sole model unknown.

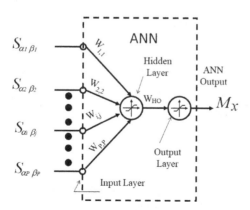

Figure 2.3. FFANN corresponding to the classical isotropic vector Preisach model.

Normalized experimental data acquired for a floppy disk sample was used to demonstrate the presented approach. For this implementation, f_{sat}, P and N were set to 1.1, 9 and 30, respectively (please refer to [37]). Two measurement sets comprised of a total of 706 $\bar{H} - \bar{M}$ pairs were used in the training (i.e., identification) phase. In specific, a set of scalar first-order-reversal curves a set of measurements correlating mutually orthogonal field and magnetization values [6]. In the later set of measurements, the normalized field is first restricted to increase along the y-axis, starting from -1.0 (i.e., $-Hsat$) up to some value Hy then back to zero thus resulting in some residual magnetization. The field is then increased along the x-axis to $+1.0$ (i.e., $+Hsat$) while monitoring both Mx and My components. The set of measurements is formed by repeating the afore-mentioned sequence for different Hy values. The training sets were fed to the FFANNs repeatedly until acceptable error margin was achieved for both networks. In this case, $Err \leq 0.5$ was considered satisfactory and the whole training process was executed within minutes using a standard personal computer.

Sample trend of the FFANN outputs after the training process are given in Fig. 2.4. Detailed comparative results between the measured and computed training data sets are given in [37]. This comparison demonstrates the qualitative agreement between measured and computed values. Although the quantitative agreement may be regarded good enough for some applications, the accuracy may be increased by

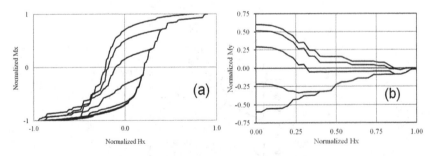

Figure 2.4. Sample trend of the FFANN outputs and measured training sets upon completion of the training process; (a) scalar output, and (b) orthogonal correlation between input and output for different residual magnetizations.

employing more primitive hysteresis operators and/or scalar models (i.e., by increasing P and/or N).

Upon termination of the identification stage, the proposed FFANN approach may be used to simulate any $\bar{H} - \bar{M}$ variations. The proposed technique is especially appealing since it uses FFANN as a tool to avoid mathematically solving the identification problem of vector Preisach models using any arbitrary available experimental data. This, undoubtedly, widens the scope of applicability for vector Preisach-type models.

2.2. Efficient Implementation of Vector Preisach Models Using Orthogonally Coupled Hysteresis Operators

It has been shown in Section 1.5 of this book that the rectangular hysteresis operator may be realized using a dual-node discrete Hopfield neural network (DHNN) as demonstrated in Fig. 1.9. Extending the above-mentioned implementation, it can be shown that the four-node DHNN shown in Fig. 2.5 may be used to realize a couple of hysteresis

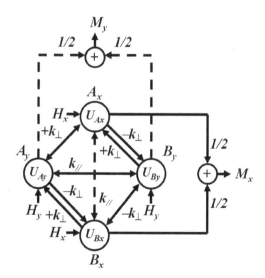

Figure 2.5. A four-node discrete Hopfield neural network (DHNN) capable of realizing two elementary hysteresis operators corresponding to the x- and y-axes.

operators whose inputs and outputs correspond to the x- and y-components [41]. More specifically, the outputs of nodes A_x and B_x can mimic the output of an elementary hysteresis operator whose input and output coincide with the x-axis. Likewise, outputs of nodes A_y and B_y can represent the output of an elementary hysteresis operator whose input and output coincide with the y-axis. In this figure, the symbols $k_{//}$ and $k_{\perp}$ denote the feedback factors between nodes corresponding to same and different axes, respectively.

According to DHNN implicit mechanism, the state of the network given in Fig. 2.5 converges to the minimum of the energy function E given by:

$$E = -\left\{ \begin{array}{l} H_x\left(U_{Ax} + U_{Bx}\right) + H_y\left(U_{Ay} + U_{By}\right) + k_{//}U_{Ax}U_{Bx} + k_{//}U_{Ay}U_{By} \\ + \dfrac{k_{\perp}}{2}\left(U_{Ax} - U_{Bx}\right)\left(U_{Ay} + U_{By}\right) + \dfrac{k_{\perp}}{2}\left(U_{Ay} - U_{By}\right)\left(U_{Ax} + U_{Bx}\right) \end{array} \right\}$$

(2.19)

Moreover, the outputs of the four DHNN nodes are updated in accordance with gradient descent rule as given by the expressions:

$$U_{Ax}(t+1) = \operatorname{sgn}\left(+k_{\perp}\left[U_{Ay}(t) + U_{By}(t)\right] + k_{//}U_{Bx}(t) + H_x\right) \quad (2.20)$$

$$U_{Bx}(t+1) = \operatorname{sgn}\left(-k_{\perp}\left[U_{Ay}(t) + U_{By}(t)\right] + k_{//}U_{Ax}(t) + H_x\right) \quad (2.21)$$

$$U_{Ay}(t+1) = \operatorname{sgn}\left(+k_{\perp}\left[U_{Ax}(t) + U_{Bx}(t)\right] + k_{//}U_{By}(t) + H_y\right) \quad (2.22)$$

$$U_{By}(t+1) = \operatorname{sgn}\left(-k_{\perp}\left[U_{Ax}(t) + U_{Bx}(t)\right] + k_{//}U_{Ay}(t) + H_y\right) \quad (2.23)$$

It should be mentioned here that the proposed implementation quali-tatively satisfies expected vector hysteresis properties previously listed in Section 1.2. More specifically, when the input is restricted along a single direction (i.e., when either H_x or H_y is equal to zero) the input-output behavior is expected to be identical to the loop shown in Fig. 1.10 with the exception that k is replaced by $k_{//}$.

The far-reaching features of this implementation concept, however, are clarified through vector-type input variations. For instance, it is possible to quantitatively modify the rotational input-output variations for the same $k_{//}$ value by simply changing $k_\perp$. This is clearly illustrated in Fig. 2.6. Samples of the mutually orthogonal input-output variations computed by the DHNN implementation for $k_{//} = 0.6$ and different $k_\perp$ values are given in Fig. 2.7. In this figure, the output remanent y-components (M_y), achieved by increasing the y-component input (H_y) to unity then back to zero, are plotted versus an increasing x-component input (H_x). As previously stated, qualitative agreement with well-known hysteresis-type vector behavior is once more demonstrated.

It is worth mentioning here that when comparable positive values of the feedback factors $k_{//}$ and $k_\perp$ are chosen, a relatively large input along a certain direction will result in the wiping out (demagnetization) of any residual output along the orthogonal direction. While a vector-type behavior observed from one DHNN block is not expected to be perfectly isotropic, an ensemble of DHNN blocks having different switching, feedback and offset values would significantly increase the overall isotropic nature of the constructed model. Computational efficiency may thus be achieved by considering a couple of scalar Preisach-type models having coupled DHNN blocks rather than the typical vector Preisach model implementation involving a collection of many scalar models oriented along all possible directions.

One way to construct a vector hysteresis model using an ensemble of the aforementioned DHNN blocks is given in Fig. 2.8. This configuration is, basically, a modular combination of Discrete Hopfield Neural Network blocks via a Linear Neural Network structure (i.e., DHNN-LNN).

In this figure, Q_i and μ_i represent the applied offset and a density value corresponding to the ith DHNN block, respectively. Obviously, Q_i and $k_{//i}$ values corresponding to the different DHNN blocks should cover the complete $\alpha - \beta$ plane region corresponding to the loop under consideration. Thus, for the DHNN #i switching up and down thresholds are α_i and β_i may be correlated to Q_i and $k_{//i}$ according to:

$$Q_i = -\left(\frac{\alpha_i + \beta_i}{2}\right) \qquad (2.24)$$

$$k_{//i} = \left(\frac{\alpha_i - \beta_i}{2} \right) \tag{2.25}$$

Please note that varying the value of $k_{\perp i}$, corresponding to DHNN #*i*, does not affect the scalar behavior of such model implementation.

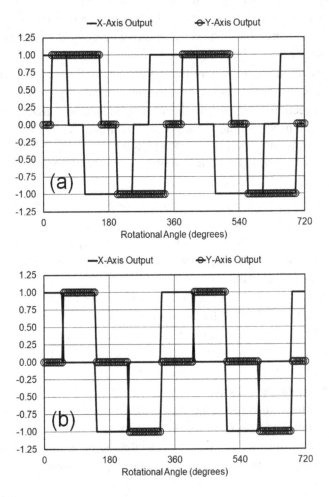

Figure 2.6. Rotational input-output variations for the DHNN shown in Fig. 2.5 corresponding to $k_{//} = 0.6$ and $k_{\perp}$ values of: (a) 0.5, and (b) 0.7.

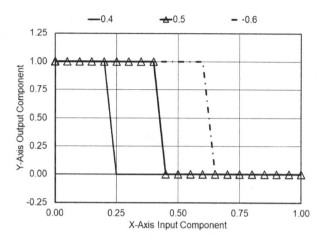

Figure 2.7. Variation of the DHNN remanent y-component outputs as a result of increasing the x-component input for $k_{//} = 0.6$ and different $k_{\perp}$ values.

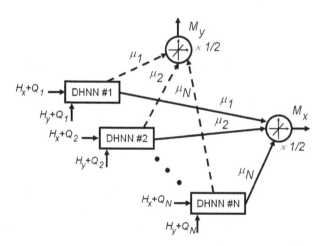

Figure 2.8. Implementation of the vector Preisach-type model using a modular combination of the DHNN blocks via an LNN.

By considering a finite number N of elementary operators, the modular DHNN-LNN shown in Fig. 2.8 evolves due to any applied input by changing output states of the operator blocks. Based upon expression (2.19), the network should eventually converge to a minimum of the quadratic energy function E given by:

$$
E = -\sum_{i=1}^{N} \left\{
\begin{array}{l}
\left[\left(H_x - \left[\dfrac{\alpha_i + \beta_i}{2} \right] \right) (U_{Axi} + U_{Bxi}) \right. \\[2ex]
+ \left(H_y - \left[\dfrac{\alpha_i + \beta_i}{2} \right] \right) (U_{Ayi} + U_{Byi}) \\[2ex]
+ \left[\dfrac{\alpha_i - \beta_i}{2} \right] U_{Axi} U_{Bxi} + \left[\dfrac{\alpha_i - \beta_i}{2} \right] U_{Ayi} U_{Byi} \\[2ex]
+ \dfrac{k_\perp}{2} (U_{Axi} - U_{Bxi})(U_{Ayi} + U_{Byi}) \\[2ex]
+ \dfrac{k_\perp}{2} (U_{Ayi} - U_{Byi})(U_{Axi} + U_{Bxi})
\end{array}
\right\}
\tag{2.26}
$$

The output vector, on the other hand, may be expressed as:

$$
M_x \bar{e}_x + M_y \bar{e}_y = \sum_{i=1}^{N} \mu_i \left[\left(\frac{U_{Axi} + U_{Bxi}}{2} \right) \bar{e}_x + \left(\frac{U_{Ayi} + U_{Byi}}{2} \right) \bar{e}_y \right]
\tag{2.27}
$$

where, $\bar{e}_x$ and $\bar{e}_y$ are unit vectors along the x- and y-directions, respectively.

The model identification for the implementation under consideration reduces to the determination of the various unknown μ_i and $k_{\perp i}$ values that would result in the best possible fit to experimental data corresponding to the specific material. Because the model is now being realized by a DHNN-LNN configuration, it is possible to carry out the vector model identification process using an automated training algorithm. Two advantages are gained in this case. First, mathematical complications embedded in the analytical identification of such a model are avoided. Second, the training process may be carried out using any available set of scalar and vector data without being restricted to a complete specific set of experimental results.

One approach to carry out the identification process is to first assume some $k_{\perp i}/k_{//i}$ ratio then pursue the determination of the unknowns μ_i for $i = 1, 2, ..., N$. This could be achieved through training using the available scalar data provided to the network and applying the least-mean-square (LMS) algorithm to the linear neuron corresponding to the output M_x. In this training process context, the error signal may be expressed in the form:

$$e(s) = M\_train_x(s) - \left[\frac{U_{Ax} + U_{Bx}}{2}\right]^T (s)[\mu](s) \qquad (2.28)$$

where,

$$[\mu] = \begin{bmatrix} \mu_1 & \mu_2 & \bullet\bullet & \mu_N \end{bmatrix}^T \qquad (2.29)$$

$$\left[\frac{U_{Ax} + U_{Bx}}{2}\right] = \left[\frac{U_{Ax1} + U_{Bx1}}{2} \quad \frac{U_{Ax2} + U_{Bx2}}{2} \quad \bullet\bullet \quad \frac{U_{AxN} + U_{BxN}}{2}\right]^T \qquad (2.30)$$

while, M_train_x represents the training x-axis data where the variable s denotes values corresponding to the sth input-output training pair.

It turns out that the LMS algorithm may be regarded as an instantaneous approximation to the steepest descent method. It is often referred to as a "stochastic gradient algorithm" [33]. Treating the instantaneous quantity $0.5 \, e^2(s)$ as a "cost function" and using a gradient descent rule for the derivative of this cost function with respect to the $[\mu]$ vector, the LMS algorithm may hence be expressed in the form:

$$[\mu](s+1) = [\mu](s) + \eta\left[\frac{U_{Ax} + U_{Bx}}{2}\right](s)e(s) \qquad (2.31)$$

where, η represents the network learning rate.

Usually, η is assigned a small value. This results in a more accurate slowly progresses adaptive process where more of the past data is remembered by the LMS algorithm. Once the scalar data training process is successfully terminated, utilization of the available vector training data may be carried out by comparing such data to corresponding model com-putations for the assumed $k_{\perp i}/k_{//i}$ ratios. By repeating the identification process after appropriately modifying the assumed $k_{\perp i}/k_{//i}$ ratios, a better match to the overall scalar and vector training data should be achieved.

Experimental testing of the validity of proposed implementation has been reported in [41]. In this testing process about 1800 DHNN blocks, comprising uniformly distributed operators over the normalized $\alpha - \beta$ region, were used. Identification (i.e., training) was initiated by setting the μ_i values to zero and the algorithm was supplied with a set of first-order reversal curves consisting of almost 1000 input-output pairs representing the scalar training data. The μ_i values were updated in accordance with (2.31) and the whole training cycle was repeated about 400 times until an acceptable mean square error value in the range of 0.01 was achieved. This training process reported in [41] was repeated for different $k_{\perp i}/k_{//i}$ ratios ranging from 0.6 to 1.2. Determination of the appropriate $k_{\perp i}/k_{//i}$ ratio that would best fit the vector data was accomplished using two sets of measurements relevant to the properties discussed in the first part of this Chapter. More specifically $\bar{M} - \bar{H}$ measurements corresponding to the application of rotational fields having different magnitudes as well as measurements corresponding to the correlation between mutually orthogonal $\bar{M}$ and $\bar{H}$ components were utilized for this purpose. For the specific magnetic media under consideration, it was inferred that optimum matching results would be achieved for a $k_{\perp i}/k_{//i}$ ratio of 1.15 [41].

Sample of the scalar and vector results reported in [41] and corresponding to the aforementioned ratios are given in Fig. 2.9. Overall qualitative and quantitative results clearly suggest that formulating the vector Preisach-type model in terms of the proposed modular DHNN-LNN combination can result in a significant reduction in computational resources required while facilitating the model identification procedure. Efficiency of this implementation stems from the fact that the vector Preisach-type model can be constructed using only two scalar models whose elementary hysteresis operators are orthogonally inter-related. Moreover, the identification problem for such an implementation may be carried out in an automated manner using well-established neural net-work algorithms.

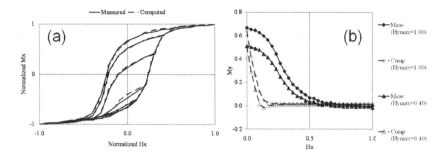

Figure 2.9. Sample (a) scalar training measured and computed curves, and (b) measured and computed orthogonally correlated Hx-My data at the end of the identification process.

A variation of the previously discussed vector hysteresis modeling methodology to extend its applicability to anisotropic media was also introduced in [42]. Assuming that the magnetization easy axis coincides with x-axis direction, the model configuration previously suggested in Fig. 2.5 is modified as shown in Fig. 2.10. In this modified configuration symbols k_e and k_h represent the feedback factors corresponding to easy-axis and hard-axis nodes, respectively.

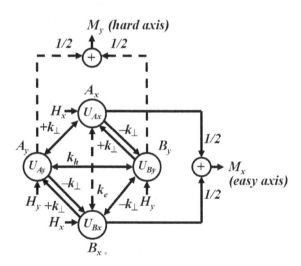

Figure 2.10. A four-node DHNN capable of realizing two elementary hysteresis operators corresponding to the easy- and hard-axes of a magnetic material.

In agreement with the previously discussed physical considerations, the proposed DHNN representation shown in Fig. 2.10 will yield outputs corresponding to a minimum value for the energy E given by:

$$E = -\left\{ \begin{array}{l} H_x\left(U_{Ax} + U_{Bx}\right) + H_y\left(U_{Ay} + U_{By}\right) + k_e U_{Ax} U_{Bx} + k_h U_{Ay} U_{By} \\ + \dfrac{k_\perp}{2}\left(U_{Ax} - U_{Bx}\right)\left(U_{Ay} + U_{By}\right) + \dfrac{k_\perp}{2}\left(U_{Ay} - U_{By}\right)\left(U_{Ax} + U_{Bx}\right) \end{array} \right\}$$

$$(2.32)$$

An important feature of this implementation may be clearly highlighted when considering variations corresponding to applied scalar and rotational input sequences. This is clearly demonstrated in Figs. 2.11 and 2.12. It should be pointed out here that the effects of modifying the feedback factor between nodes corresponding to orthogonal axes are also shown in Fig. 2.14.

Just like for the case of the implementation of isotropic vector models, the anisotropic vector hysteresis model may be realized by weeding an ensemble of the newly proposed DHNNs via a network similar to that shown in Fig. 2.8. Within this implementation, the model identification is once more reduced to the determination of the μ_{xi}, μ_{yi} and $k_\perp$ values that would best fit the given identification (training)

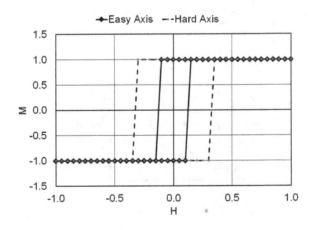

Figure 2.11. DHNN outputs corresponding to scalar easy- and hard-axes inputs for $k_e = 0.1$ and $k_h = 0.3$.

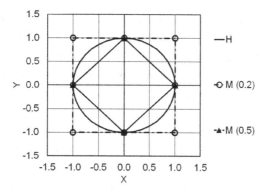

Figure 2.12. Computed magnetization resulting from applying a unity rotational input on a DHNN having $k_e = 0.1$ and $k_h = 0.3$ corresponding to $k_\perp$ values of 0.2 and 0.5.

experimental data. As previously stated, the unknowns of the DHNN-LNN configuration (i.e., the identification process) may be using standard automated algorithms for any available set of scalar and vector data. Unlike the case of isotropic models, matching scalar data here involves both easy and hard axes data. This identification process may be carried out by first assuming some $k_\perp$ value then carrying out the network training process using the scalar easy and hard axes data. Again, the stochastic gradient algorithm may be applied to the linear neurons whose outputs represents M_x and M_y [33]. During this process, the identification error for both axes corresponding to the training input-output pair number "s" may be given by:

$$\begin{bmatrix} e_x(s) \\ e_y(s) \end{bmatrix} = \begin{bmatrix} M_{tx}(s) \\ M_{ty}(s) \end{bmatrix} - \sum_{i=1}^{N} \begin{bmatrix} \mu_{xi}(s)\left(\dfrac{U_{Axi}(s)+U_{Bxi}(s)}{2}\right) \\ \mu_{yi}(s)\left(\dfrac{U_{Ayi}(s)+U_{Byi}(s)}{2}\right) \end{bmatrix} \quad (2.33)$$

In (2.33), the unknown weights μ_{xi} and μ_{yi} are updated according to:

$$\begin{bmatrix} \mu_{xi}(s+1) \\ \mu_{yi}(s+1) \end{bmatrix} = \begin{bmatrix} \mu_{xi}(s) \\ \mu_{yi}(s) \end{bmatrix} + \eta \begin{bmatrix} e_x(s)\left(U_{Axi}+U_{Bxi}\right)/2 \\ e_y(s)\left(U_{Ayi}+U_{Byi}\right)/2 \end{bmatrix} \quad (2.34)$$

where, η is some learning rate (refer, for instance, to [33]).

Once the scalar training process is finished by achieving an acceptable error margin, available identification vector data may be then utilized by comparing measured and computed vector training data for different pre-set $k_\perp$ values. Hence, an appropriate value for $k_\perp$ that gives the best data match with vector training (identification) data may be determined.

Testing of the proposed approach has been reported in [42]. More specifically, experimental data for Fe-Si sample was utilized as the easy axis scalar data. Hard axis scalar data was generated by assuming demagnetization factors. In other words, the testing involved shape introduced anisotropy rather than inherent material anisotropy. In the reported numerical implementation, about 6500 DHNN blocks which are uniformly distributed over the normalized limiting H-M curve were used. Comparison between measured and computed scalar data by the end of the training (identification process) is given in [42]. Samples of the computed results are given in Fig. 2.13. It should be mentioned here that the extent of quantitative match between measured and computed scalar data is, more or less, independent of the $k_\perp$ value. Additional results demonstrating the model capability to match different rotational data are given in Fig. 2.14. From a qualitative viewpoint, these results are in full agreement with similar reported measurement (see, for instance, [43]).

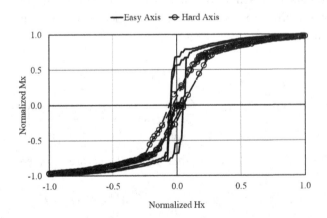

Figure 2.13. Sample computed normalized scalar data towards the end of the training process for the Fe-Si sample.

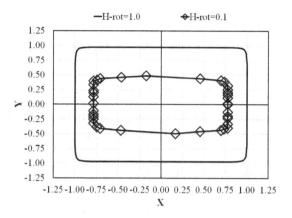

Figure 2.14. Sample normalized Fe-Si vectorial output simulation results corresponding to a rotational applied input having a normalized amplitude of 1.0 and 0.1.

2.3. Efficient Implementation of Vector Preisach Models Using Orthogonally Coupled Stoner-Wohlfarth-Like Hysteresis Operators

It has been shown in Section 1.6 that a Stoner-Wohlfarth-like hysteresis operator may be realized using a dual node HNN having hybrid discrete-continuous activation functions and positive feedback (see, for instance, [15] and [34]). By utilizing a four-node HNN similar to that shown Fig. 2.5 and having hybrid discrete-continuous activation functions, the network state may be deduced by minimizing the energy function E given by expression (2.19). In this case, node outputs are updated in accordance with the following expressions:

$$U_{Ax}(t+1) = cfc(net_{Ax}(t)) + dfd(net_{Ax}(t)),$$
$$net_{Ax}(t) = Hx + k_{//}U_{Bx}(t) + k_{\perp}\left(U_{Ay}(t) + U_{By}(t)\right)$$
(2.35)

$$U_{Bx}(t+1) = cfc(net_{Bx}(t)) + dfd(net_{Bx}(t)),$$
$$net_{Bx}(t) = Hx + k_{//}U_{Ax}(t) - k_{\perp}\left(U_{Ay}(t) + U_{By}(t)\right)$$
(2.36)

$$U_{Ay}(t+1) = cfc(net_{Ay}(t)) + dfd(net_{Ay}(t)),$$
$$net_{Ay}(t) = Hy + k_{//}U_{By}(t) + k_{\perp}\left(U_{Ax}(t) + U_{Bx}(t)\right)$$
(2.37)

$$U_{By}(t+1) = cfc(net_{By}(t)) + dfd(net_{By}(t)),$$

$$net_{By}(t) = Hy + k_{//}U_{Ay}(t) - k_{\perp}\left(U_{Ax}(t) + U_{Bx}(t)\right)$$
(2.38)

where,

$$d = 1-c, \quad 0 \le c \le 1$$
(2.39)

Restricting the input to vary along a single direction will result in *M-H* curves similar to those shown in Fig. 1.12. On the other hand, the far-reaching capabilities of the proposed four-node hybrid HNN may be demonstrated when rotational and mutually orthogonal input-output variations are considered. Sample results related to the aforementioned input-output vector variations are given in Figs. 2.15 and 2.16. These figures clearly suggest that the proposed implementation qualitatively satisfy experimentally observed vector hysteresis properties. Building upon the reasoning mentioned in Section 1.6, utilizing the Stoner-Wohlfarth-like operators in the implementation will result in a computationally efficient vector hysteresis model comprised of a reduced number of HNN blocks interconnected as shown in Fig. 2.8.

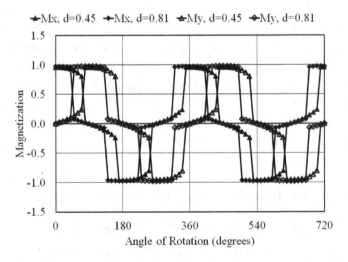

Figure 2.15. Mx and My components of the proposed four-node HNN resulting from applying a rotational input of unity amplitude and corresponding to $k_{//}$ = 0.48/d and $k_{\perp}$ = 0.3.

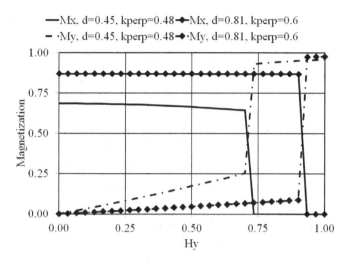

Figure 2.16. Correlation between mutually orthogonal input-output components for the proposed four-node HNN corresponding to $k_{//} = 0.48/d$ (N.B. $k_\perp$ is denoted by kperp).

For the modular model representation depicted in Fig. 2.8, the offset values Q_i and positive feedback factors $k_{//}$ corresponding to the different Stoner-Wohlfarth-like smooth operators are chosen to cover the hysteretic loop zone. Moreover, varying the orthogonal coupling factor $k_\perp$ would mainly affect the vector performance of an HNN blocks. For a finite number N of the proposed HNN blocks the identification of the model unknowns is reduced to the proper selection of unknowns c (and consequently d), $k_\perp$ and HNN block density values μ_i. Initiating the identification process with preset values of c and $k_\perp$ the modular HNN evolves, as a result of applied inputs, by changing its output states to satisfy a minimized value of the of energy function E given by expression (2.26). HNN outputs may, this, be computed from expression (2.27).

Appropriate μ_i values are determined during this identification (training) phase using the available scalar data provided to the network and the least-mean-square (LMS) algorithm implicitly adopted in the LNN neuron whose output corresponds to Mx. Since c is closely related to the hysteresis loop squareness, the training process is repeated to identify the optimum value of this parameter that would lead to the minimum matching error with the available scalar data. Once the scalar data training process is completed, the available vector training data may then be employed to determine the optimum $k_\perp$ value.

To demonstrate the efficiency and accuracy of the proposed approach simulations and experimental testing were demonstrated in [15] for a floppy disk sample. Given the nature of the Stoner-Wohlfarth-like operators, it was only necessary to utilize HNN blocks restricted to the normalized coercive field range of the major loop. As a result, about 25% of the total number of HNNs utilized in the previously proposed approaches of Section 2.2 were needed. This fact clearly highlights the computational efficiency of the approach. Moreover, since μ_i values corresponding to operators whose switching values are symmetric with respect to the $\alpha = -\beta$ line should be the same as explained in [6], unknown block density values were reduced to half of the total number of blocks (about 200 in this case). During the identification (training) phase a set of first-order reversal curves comprised of about 1000 *Hx-Mx*

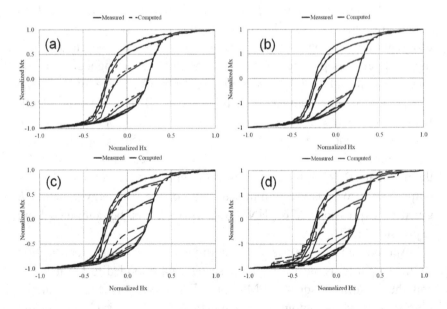

Figure 2.17. Comparison between measured and computed first-order-reversal curves towards the end of the training process for; (a) d = 0.1, $k_\perp$ = 0.6, (a) d = 0.2, $k_\perp$ = 0.6, (a) d = 0.1, $k_\perp$ = 0.8, and (b) d = 0.1, $k_\perp$ = 1.1.

pairs was used. The training cycle was repeated for pre-set values of c and $k_\perp$ until the mean square error reached an acceptable value (in the order of 0.01). Sample results reflecting the outcome of the training process for different c and $k_\perp$ values are shown in Fig. 2.17.

Assessment of the most appropriate $k_\perp$ value was carried out by comparing computed and measured results related to the application of rotational field values having different amplitudes (see Fig. 2.18). It should be mentioned that for the sample under consideration reported optimum values for c and $k_\perp$ were found to be equivalent to 0.5 and 0.1, respectively [15]. In order to test the model ability to predict measurements not included in the identification process, comparisons were carried out for a set of data related to the correlation property of mutually orthogonal components. Results of these comparisons, which were presented in detail in [15], clearly highlight the model prediction accuracy.

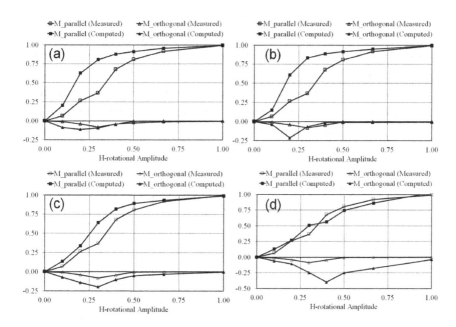

Figure 2.18. Comparison between measured and computed rotational data corresponding to; (a) d = 0.1, $k_\perp$ = 0.6, (a) d = 0.2, $k_\perp$ = 0.6, (a) d = 0.1, $k_\perp$ = 0.8, and (b) d = 0.1, $k_\perp$ = 1.1d = 0.5 for; (a) $k_\perp$ = 0.8, and (b) $k_\perp$ = 1.0.

2.4. Vector Hysteresis Modeling Using Rotationally Coupled Step Functions

Another different computationally efficient approach for implementing vector hysteresis models was introduced in [44] and [45]. In this approach, which may be utilized to model isotropic and anisotropic media, an ensemble of octal clusters of coupled step functions is utilized as given in Fig. 2.19. Consider the eight-node Discrete Hopfield Neural Network (DHNN) block shown in Fig. 2.19.a where the activation function output of each node is given by $U \in [-1, +1]$. The overall energy E of this DHNN may be given by:

$$E = -\bar{H} \bullet \sum_{i=1}^{8} U_i \bar{e}_i - k_{ij} \sum_{i=1}^{8} \sum_{\substack{j=1 \\ j \neq i}}^{8} \left(U_i \bar{e}_i \bullet U_j \bar{e}_j \right), \text{ and } k_{ij} = \begin{cases} -k_{//} & \text{for } \bar{e}_i \bullet \bar{e}_j = -1 \\ +k_c & \text{otherwise} \end{cases}$$

(2.40)

where, $\bar{H}$ is the applied field, $k_{//}$ is the self-coupling factor between any two step functions having opposite orientations, k_c is the mutual coupling factor, while U_i is the output of the ith step function oriented along the unit vector $\bar{e}_i$.

For the DHNN shown in Fig. 2.19.a each node applies a step function to the summation of the applied vector field projection and the weighted output states of all other nodes in the network. The DHNN

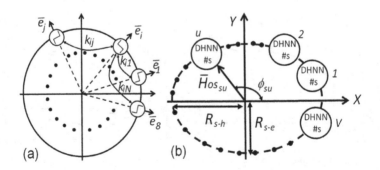

Figure 2.19. (a) Coupled eight-node DHNN having step activation functions, and (b) Generally (circularly or elliptically) dispersed ensemble of V similar DHNN blocks shown in (a).

internally updates the output states corresponding to a fixed external field for its various nodes until the network overall state converges to the minimum overall energy E expressed as given by expression (2.40) [33]. By tuning the values of $k_{//}$ and k_c the scalar and vectorial performance of the DHNN under consideration may be varied. This is clearly illustrated by Figs. 2.20 and 2.21.

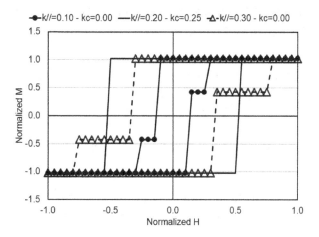

Figure 2.20. Effect of tuning $k_{//}$ and k_c on the scalar output of the eight-node DHNN.

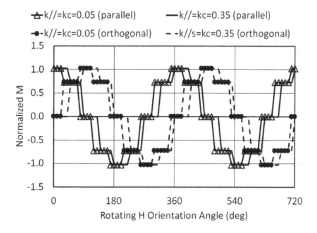

Figure 2.21. Effect of tuning $k_{//}$ and k_c on the rotational output of the eight-node DHNN.

It may thus be inferred that a computationally efficient hysteresis model may be constructed using a limited vectorially dispersed ensemble of the previously discussed DHNN blocks. This vectorial dispersion may take a circular form to represent isotropic media or, alternatively, elliptical form dispersion as given in Fig. 2.19.b to extend the model applicability to anisotropic media [44], [45].

It turns out that for the rotationally dispersed ensemble denoted by "s" of V similar DHNN blocks shown in Fig. 2.19.b, the total input field applied to the uth DHNN block $\bar{H}_{tu}$ may be given by the expression:

$$\bar{H}_{tu} = \bar{H} + \bar{H}o_{su} = \bar{H} + \frac{e^{j\phi_{su}}}{\sqrt{\dfrac{\cos^2 \phi_{su}}{R_{s-e}^2} + \dfrac{\sin^2 \phi_{su}}{R_{s-h}^2}}}, \quad \phi_{su} = \frac{2\pi}{V}\left(u - \frac{1}{2}\right) \quad (2.41)$$

where, $\bar{H}o_{su}$ is an imposed vectorial field offset while R_{s-e} and R_{s-h} represent the different radii of the elliptic vectorial dispersion, respectively.

Enhancement of the quantitative accuracy of this vector hysteresis modelling methodology may be achieved by employing a superposition of vectorially dispersed ensembles of DHNN blocks having different $k_{//}$, k_c, R_{s-e} and R_{s-h} values. It should be stated here that for anisotropic media, total number of unknowns becomes four times the number of DHNN ensembles employed. For isotropic media, on the other hand this number reduces to three times since in such case $R_{s-e} = R_{s-h}$.

One way to handle the identification problem of the model is to employ the Particle Swarm Optimization (PSO) evolutionary computation technique (refer, for instance, to [41], [46] and [47]). This technique simulates the social behavior of insect swarms on a population (representing model unknowns) of possible solutions of the search space where the goal is to minimize the root mean square error ε_{RMS} between computed and measured identification data. During the identification process each PSO generated model unknown set is used to compute vector outputs corresponding to input vector set of the identification data. This PSO module is executed by using a swarm of particles, P, each keeping record of its own attributes. The most important attribute of each particle is its current position as given by an D-dimensional vector representing the model unknowns $Q^t = (Q_1^t, Q_2^t, ..., Q_D^t)$. Each particle also keeps track of its current velocity, $vp^t = (vp_1^t, vp_2^t, ..., vp_D^t)$, current fitness value ε_{RMS-p}^t and best position leading to minimum error, $p^t = (p_1^t, p_2^t, ..., p_D^t)$. The best overall position among all particles, p^g, is also recorded.

During each epoch, every particle is accelerated towards its own personal best as well as the global best position according to the following expressions:

$$vp_i^t = w \times vp_i^t + c_1 \times rand1(\ ) \times (\ p_i^t - Q_i^t) + c_2 \times rand2(\ ) \times (\ p_i^g - Q_i^t) \quad (2.42)$$

$$Q_i^t = Q_i^t + vp_i^t \quad (2.43)$$

where, $rand1()$ and $rand2()$ are random functions in the range $[0,1]$, t, c_1 and c_2 are positive constants, while w is the inertia weight [47].

Eventually, p^g serves as the algorithm solution of the model unknowns that yields minimum identification data fitting error ε_{RMS}.

In order to check the validity of this approach, numerical implementation and experimental testing have been carried out for two magnetic samples; an isotropic floppy disk sample as well as an anisotropic magnetic tape sample [45]. For every sample, the identification process was first carried out using a combination of four DHNN ensembles having $N = V = 8$ similar to those shown in Fig. 2.19. For this particular numerical implementation, the total number of model unknowns corresponding to the isotropic case amounted to 12 (i.e., $k_{//}$, k_c and R_i for each DHNN ensemble). For the anisotropic case, on the other hand, the total number of model unknowns amounted to 16 (i.e., $k_{//}$, k_c, R_{i-h} and R_{i-e} for each DHNN ensemble). It should be mentioned here that using such a configuration a vector hysteresis model may be constructed using only 132 rectangular operators which is a significantly small number compared to the case of a typical conventional vector Preisach type model. Measured sets of first-order reversals and measurements correlating orthogonal input and output components for both samples were employed by the previously mentioned PSO algorithm to identify the optimum values of the unknowns. Sample results representing the outcome of the identification process are given in Figs. 2.22 and 2.23 [45].

Additional testing was performed to check the model ability to predict data that was not involved in the identification process (please refer to [44] and [45]). Sample computed values for the parallel and orthogonal magnetization components corresponding to an applied rotational field is shown in Fig. 2.24.

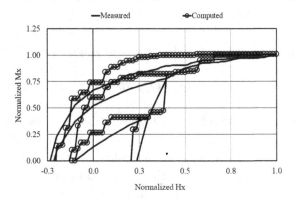

Figure 2.22. Sample comparison between measured and computed first-order-reversal curves after concluding the identification process.

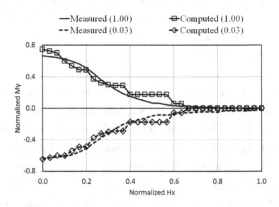

Figure 2.23. Sample comparison between measured and computed orthogonal input-output correlation curves.

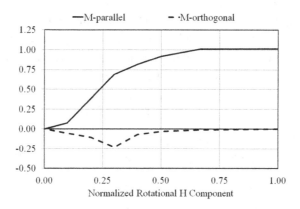

Figure 2.24. Sample computed values for the parallel and orthogonal magnetization components corresponding to an applied rotational field.

2.5. Vector Hysteresis Modeling of Bulk 2D and 3D Arbitrary Shaped Objects

It has been shown that vector hysteresis models may be constructed through the superposition of Stoner-Wohlfarth-like elementary hysteresis operators which are realized using dual node HNNs having positive feedback and hybrid discrete-continuous activation functions (refer, for instance, to [15]). Overall magnetic behavior of a bulk magnetic material may involve shape induced anisotropy (see, for example, [48] and [49]).

For 2D geometrical configurations, discretization is usually carried out via an ensemble of triangular subregions. In such cases, a newly introduced modeling approach further generalized the previously presented work by correlating an arbitrary triangular geometrical configuration to the coupling coefficients of a corresponding tri-node HNN [50]. In other words, an elementary vector hysteresis operator representing a triangular subregion has been introduced. In order to shed some light on this approach, consider a tri-node HNN having positive feedbacks and hybrid discrete-continuous activation functions. For the uth triangular subregion having arbitrary side lengths, the tri-node HNN may be represented as given in Fig. 2.25. According to this novel implementation, activation functions are located at the triangle vertices. Moreover, the input-output orientation of every node is assumed to be along the line joining the uth triangle centroid to its corresponding vertex. For any two nodes, say i and j, the orientation dependent positive feedback $k_{i,j}^{(u)}$ coupling these nodes is assumed to be given by:

$$k_{i,j}^{(u)} = k_{copl} \left| \cos(\varphi_{ij}^{(u)}) \right| \qquad (2.44)$$

where, k_{copl} is a pre-selected coupling factor ≤ 1 and $\phi_{ij}^{(u)}$ is the angle subtended between nodes i and j orientations as shown illustrated by Fig. 2.26.

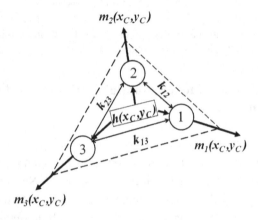

Figure 2.25. Realization of a basic vector hysteresis operator representing a triangular sub-region using a tri-node HNN.

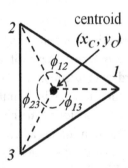

Figure 2.26. Angles between the assumed vector orientations of the various tri-node HNN activation functions.

Assuming a hybrid activation function $f(x)$ as presented in Section 1.6, the hybrid activation rule for the ith node may expressed in the form:

$$m_i^{(u)}(t+1) = c\, f_c(net_i^{(u)}(t)) + (1-c)\, f_d(net_i^{(u)}(t)) \qquad (2.45)$$

where, c is a positive constant ≤ 1 while $f_c(x)$ and $f_d(x)$ are the sigmoid continuous and signum discrete activation functions, $m_i^{(u)}$ is the node output while $net_i^{(u)}$ is given by:

$$net_i^{(u)}(t) = \bar{h}^{(u)}(t) \bullet \left(\frac{\bar{m}_i^{(u)}}{m_i^{(u)}}\right) + \sum_{j=1, j\neq i}^{3} k_{i,j}^{(u)} m_j^{(u)}(t) \qquad (2.46)$$

In (2.46), $\bar{h}^{(u)}$ represents the external field applied to the uth triangular subdivision.

For the tri-node HNN under consideration corresponding to the uth triangular subdivision, network state converges to the minimum of the energy function $E^{(u)}$ given by:

$$E^{(u)} = -\sum_{i=1}^{3}\left[\bar{h}^{(u)} \bullet \bar{m}_i^{(u)} + \frac{1}{2} \sum_{j=1, j\neq i}^{3} k_{i,j}^{(u)} m_i^{(u)} m_j^{(u)} \right] \qquad (2.47)$$

In accordance with the aforementioned equations HNN algorithms are employed to achieve a minimum value for $E^{(u)}$ corresponding to every applied external input value $\bar{h}^{(u)}$ (refer, for instance, to [20], [32] and [33]). By superposition, the overall triangle output $\bar{m}^{(u)}$ may thus be computed from:

$$\bar{m}^{(u)} = \sum_{i=1}^{3} \bar{m}_i^{(u)} \qquad (2.48)$$

The effect of the geometrical configuration on a triangle's primitive scalar and vector hysteresis properties has been demonstrated by considering three different triangular configurations denoted by Tr_1, Tr_2 and Tr_3 for a specific set of model parameters k_{copl}, a and c. Angular orientations of the three triangular subdivisions were assumed to be ($0°$, $+120°$, $-120°$), ($0°$, $+110°$, $-110°$), and ($+10°$, $+110°$, $-140°$)

respectively. A comparison between the scalar *M-H* curves along the x-axis direction as well as the rotational input-output variations corresponding to a unity rotational applied input for the three triangles are given in Figs. 2.27 and 2.28, respectively.

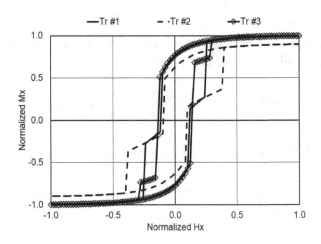

Figure 2.27. Normalized scalar M-H curves along the x-axis corresponding to the three triangles under consideration for $k_{copl} = 0.3$, $a = 3$ and $c = 0.6$.

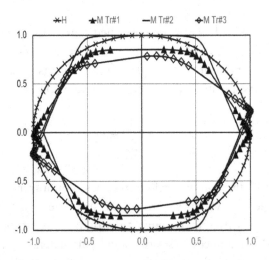

Figure 2.28. Normalized rotational M-H curves corresponding to the three triangles under consideration for $k_{copl} = 0.3$, $a = 3$ and $c = 0.6$.

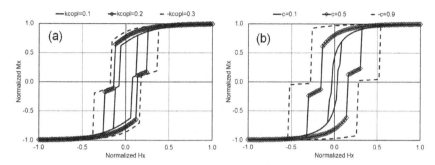

Figure 2.29. Normalized scalar M-H curves along the x-axis corresponding to the triangle Tr_1 for (a) a = 3 and c = 0.6, and (b) k_{copl} = 0.3 and a = 3.

It should be mentioned here that different values of an *M-H* curve width and squareness for a particular triangular subdivision may be realized by tuning the different model parameters. This is clearly demonstrated in Fig. 2.29 where the effect of varying k_{copl} and c on the overall *M-H* curve is highlighted. This fact suggests that the approach under consideration may be tuned to custom fit specific hysteresis curves.

It turns out that in order to fit the model to the data relevant to an isotropic media, a circular 2D geometrical shape should be considered to exclude shape induced anisotropy. In the case when the 2D shaped magnetic body is circular, no shape anisotropy is introduced. More specifically, taking a measured magnetic material *M-H* curve it was possible to tune the model unknowns to fit the hysteresis loop for a circular region approximated by an ensemble of 36 triangular subregion. The circular region as well as the result of the *M-H* curve fitting process are shown in Figs. 2.30 and 2.31, respectively. It should be mentioned that optimum model parameters, computed using the least square errors method, that led to the quantitatively good matching between measured and computed *M-H* curves, were found to be $k_{copl} = 0.38$, $a = 1.2$ and $c = 0.3$. Checking the rotational input-output correlations using the aforementioned optimum model values good qualitative and quantitative agreement between measured and computed values were also observed. These results, that clearly demonstrate the isotropic nature of the 2D body, are given in detail in [50].

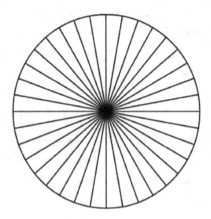

Figure 2.30. A circular 2D region approximated by an ensemble of 36 triangular sub-regions.

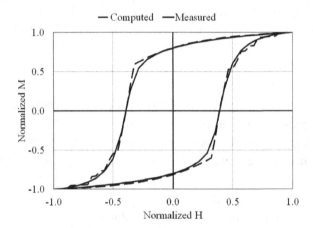

Figure 2.31. Comparison between measured and computed hysteresis loop for the circular subregion comprised of 36 triangular sub-regions shown in Fig. 2.30 corresponding to model parameters: $k_{copl} = 0.38$, $a = 1.2$ and $c = 0.3$.

The ability of the approach under consideration to account for shape induced anisotropy may be demonstrated by considering 2D geometrical configurations in the form of an ellipse, a square, and a rectangle. For these three configurations, number of triangular subdivisions was maintained at 36 as shown in Fig. 2.32 (i.e., the same number of triangular subdivisions previously employed for the circular configuration). For the elliptic and rectangular configurations, aspect ratio between easy and hard axes were taken as 2 to 1. Effect of the bulk 2D material shape induced anisotropy on the expected scalar and vector magnetization behavior while maintaining $k_{copl} = 0.3$, $a = 3$ and $c = 0.7$ is presented in Fig. 2.33. Obviously, the scalar hysteresis loop of a square object deviates from that of a circular object as can be concluded from the figures under consideration. Scalar data also reveals the effect of shape anisotropy on the variation of easy and hard axes *M-H* curves for the rectangular geometrical configuration. Moreover, the square nature of the 2D geometrical bulk object as well as the anisotropy of both the elliptical and rectangular configurations may be easily inferred from the rotational *M-H* curves shown in the same figure.

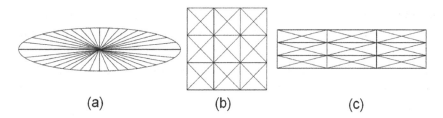

(a)   (b)   (c)

Figure 2.32. (a) Elliptic, (b) square and (c) rectangular 2D configurations, each sub-divided into triangular sub-regions.

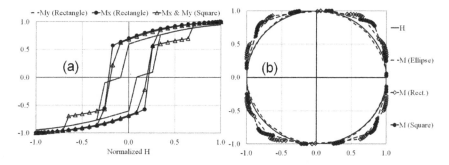

Figure 2.33. Computational results demonstrating the effect of shape induced anisotropy for the configurations shown in Fig. 2.36; (a) Scalar M-H curves along the along the x- and y-axes for the square and rectangular configurations and, (b) rotational M-H curves.

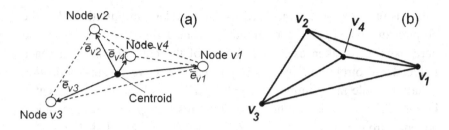

Figure 2.34. (a) The four-node HNN highlighting the assumed orientation of each neuron activation function corresponding to. (b) a typical tetrahedron.

A 3D extension of the presented methodology was also presented in [51]. The approach utilizes a four-node tetrahedron-shaped HNN with hybrid discrete-continuous activation functions similar to those employed in the 2D case. This may be clarified by considering the tetrahedron having vertices $v_1 - v_4$ and its analogous four-node HNN as shown in Fig. 2.34. As can be seen from this HNN, orientation of the different activation functions $\bar{e}_{v1} - \bar{e}_{v4}$ are assumed to be along the direction initiated from the tetrahedron centroid to the vertex of the corresponding node. Total output of the tetrahedron HNN may be deduced form the expressions:

$$\bar{m}_v(\bar{H}) = \sum_{t=1}^{4} m_{vt}(\bar{H})\bar{e}_{vt} \tag{2.49}$$

$$m_{vt} = cf_c(\omega_{vt}) + (1-c)f_d(\omega_{vt}) \tag{2.50}$$

$$f_c(\omega_{vt}) = \tanh(a\omega_{vt}), \quad a > 0 \tag{2.51}$$

$$f_d(\omega_{vt}) = \begin{cases} +1 & \text{if } \omega_{vt} > 0 \\ -1 & \text{if } \omega_{vt} < 0 \\ \text{unchanged} & \text{if } \omega_{vt} = 0 \end{cases} \tag{2.52}$$

$$\omega_{vt} = \bar{H} \bullet \bar{e}_{vt} + \sum_{r=1, r \neq t}^{4} m_{vr} k_{copl} |\bar{e}_{vr} \bullet \bar{e}_{vt}| \tag{2.53}$$

where, m_{vt}, ω_{vt}, f_c and f_d are the HNN node output, input, continuous activation function, and discrete activation function, respectively, while $\bar{e}_{vt}$ is the unit vector identifying the output orientation of HNN node t in the four-node HNN$_v$. In the previous equations, a, c and k_{copl} are, once more, the intrinsic primitive model parameters using which the overall hysteresis curve may be modified through the identification phase to fit that of a specific material.

In accordance with the typical HNN energy minimization algorithm, the implementation under consideration yields an output corresponding to a minimum state for the four-node HNN v energy E_v given by:

$$E_v = -\sum_{t=1}^{4}\left[\bar{H} \bullet m_{vt}\bar{e}_{vt} + \frac{1}{2}\sum_{r=1,r\neq t}^{4} m_{vr}m_{vt}k_{copl}\left|\bar{e}_{vr} \bullet \bar{e}_{vt}\right|\right] \quad (2.54)$$

Effects of the tetrahedron geometrical configuration variation as well as changing the $M - H$ curve model parameters may be easily observed from the scalar and rotational results shown in Figs. 2.35 and 2.36. In these figures, two hypothetical tetrahedra were considered, both sharing the three vertices' coordinates given by: (-1, 0.5, 0), (-0.5, -1, 0), and (0, 0, 1). For results shown in Fig. 2.35.a, 2.35.b and 2.36.a the fourth vertex v_4 was assumed to be (4, 0, 0). On the other hand, vertex for Figs. 2.35.c, and 2.36.b was assumed to be (2, 0, 0). Model parameter c was varied while maintaining parameters a and k_{copl} at 3 and 0.2, respectively.

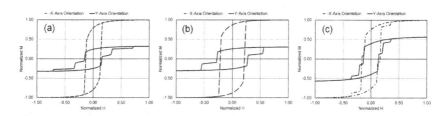

Figure 2.35. Normalized x and y input-output variations for (a) c = 0.3 and v_4 = (4, 0, 0), (b) c = 0.5 and v_4 = (4, 0, 0), and (c) c = 0.5 and v_4 = (2, 0, 0).

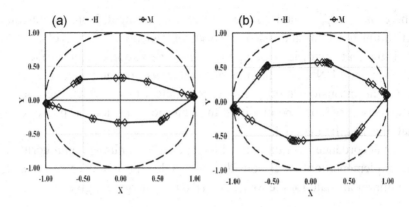

Figure 2.36. Normalized cyclic input-output variations corresponding to: (a) $v_4 = (4, 0, 0)$ and (b) $v_4 = (2, 0, 0)$.

Given that any 3D bulk object may be approximately constructed using an ensemble of tetrahedra, the methodology under consideration may be employed to assess overall behavior of magnetic bodies having arbitrary geometrical configurations. Sample assessment of bulk cylindrical magnetic objects corresponding to two model parameters and having different aspect ratios (AR) are shown in Figs. 2.37 and 2.38. More specifically, model parameter c was varied while maintaining parameters a and k_{copl} at 3 and 0.2, respectively. Moreover, cylinder height-to-diameter aspect ratios of 1.0, 1.5 and 2.0 were considered. Throughout the computations, each cylindrical object was approximated by an ensemble of 288 tetrahedra. Total magnetization $\bar{M}$ of the 3D cylindrical object 3D was, thus, computed by superposition of the various tetrahedron magnetizations in accordance with:

$$\bar{M}(\bar{H}) = \sum_{v=1}^{288} \bar{m}_v(\bar{H}) \qquad (2.55)$$

Finally, it should be pointed out that throughout this Chapter different unconventional approaches to model vector magnetic hysteresis were presented. Computational efficiency of these models was stressed as a result of either the relative sophistication of the primitive hysteresis operator and/or the intercoupling between operators corresponding to different orientations. It was, thus, obvious to capitalize on this computational efficiency by incorporating these models in field computational approaches in media exhibiting nonlinear magnetic media and/or media exhibiting magnetic hysteresis. Examples of such approaches may be referred to in [34], [51], [52], [53] and, [54].

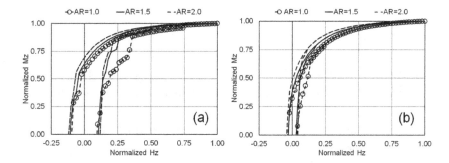

Figure 2.37. Normalized M-H curves along the cylinder axis corresponding to different height-to-diameter aspect ratios for (a) c = 0.3 and, (b) c = 0.1.

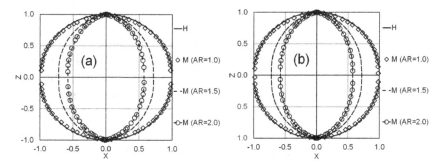

Figure 2.38. Normalized rotational field-magnetization curves along the x-z plane corresponding to different height-to-diameter aspect ratios for (a) c = 0.3 and, (b) c = 0.1.

Chapter 3

Magnetic Hysteresis Models for Field-Temperature Variations

Prediction of temperature variation effects on the magnetic properties is crucial for estimating the operation performance as well as the reliability of a wide range of spectrum of devices utilizing these materials (refer, for instance, to [20] and [55]). Examples include those related to magnetic recording, electrical machines, sensors, and actuators. As previously discussed in Section 1.2, temperature variation could lead to an irreversible change in the magnetization of a magnetic material in addition to a modification of its *M-H* curve specifics, especially the coercivity. For the case when one dimensional *M-H* variations are considered temperature changes may qualitatively lead to effects similar to those resulting from the application of an applied orthogonal magnetic field. In view of this fact, the notion of employing a two-dimensional vector Preisach hysteresis model to simulate magnetic media exhibiting hysteresis subject temperature variations and one-dimensional applied field and temperature variations was first introduced in [56]. Precisely, the x-axis normalized magnetization *M* was correlated to the applied x-axis normalized magnetic field *H* and normalized temperature *T* by designating them as the two-dimensional model x-axis output, x-axis input and y-axis input, respectively. The model under consideration may, thus, be expressed in the form:

$$M(t) = \int\limits_{-\pi/2}^{+\pi/2} \cos\varphi \left[\iint\limits_{\alpha \geq \beta} \mu(\alpha,\beta) f(\varphi) \hat{\gamma}_{\alpha\beta} \left(\overline{e}_\varphi \cdot \begin{bmatrix} H(t)\,\overline{e}_x \\ +T(t)\,\overline{e}_y \end{bmatrix} \right) d\alpha \, d\beta \right] d\varphi$$

(3.1)

where, t is the input-output time instant, $\hat{\gamma}_{\alpha\beta}$ are elementary rectangular hysteresis operators with α and β being the up and down switching values as shown in Fig. 1.3, $\bar{e}_{\varphi}$ is a unit vector along the polar angle φ, f is some pre-chosen even function, while μ is the model unknown that has to be determined through the identification procedure.

By interchanging the order of integration for expression (3.1) and assuming a finite number of elementary operators we get:

$$M(t) = \sum_{\alpha_i \geq \beta_j} \mu(\alpha_i, \beta_j)\, \mathrm{P}_{\alpha_i \beta_j} \qquad (3.2)$$

where,

$$\mathrm{P}_{\alpha_i \beta_j} = \int_{-\pi/2}^{+\pi/2} \cos\varphi\, f(\varphi)\hat{\gamma}_{\alpha_i \beta_j} \left(\bar{e}_{\varphi} \cdot \left[H(t)\, \bar{e}_x + T(t)\, \bar{e}_y \right] \right) d\varphi \qquad (3.3)$$

Given the different advantages of utilizing artificial neural networks (ANNs) in constructing vector hysteresis models, which were discussed in Chapter 2, the same methodologies may be extended to cover simulations of field-temperature effects on magnetic media [57]. More specifically, the model under consideration may be realized by the ANN shown in Fig. 3.1. This two-layer ANN consists of an input stage, one hidden layer, and an output layer of neurons successively connected in a feed-forward fashion. Number of nodes in the input layer are equivalent to the number of elementary hysteresis operators necessary to achieve a model with acceptable accuracy. On the other hand, each of the hidden and output layers contains a single neuron. The activation function employed in the ANN is bipolar sigmoidal activation function f_{sig} given by:

$$f_{sig} = -1 + \frac{2}{(1 + e^{-x})} \qquad (3.4)$$

For a pre-assumed $f(\varphi)$, the identification process of the model under consideration is carried out using a set of measured first order M-H reversal curves corresponding to different T values. The integration $P_{\alpha\beta}$ given by (3.3) is evaluated at every time instant t for every hysteresis

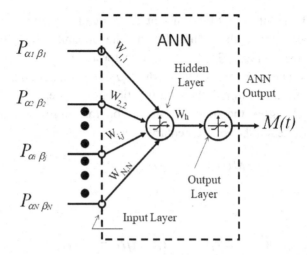

Figure 3.1. Realization of the H-T-M model using a feed forward ANN.

operator identified by its specific switching values (α_i, β_j), thus leading to the generation of the training input-output data required by the ANN to determine its optimum branch weights $W_{i,j}$ and W_h. Training is implemented repeatedly, using the ANN back-propagation supervised learning algorithm. Training repeatedly modifies the branch weights, starting from some random values (refer, for instance, to [33] and [32]). During this training process, minimization of the cumulative mean squared error between the training measured and actual network outputs is carried out. The mean squared error is minimized based on the gradient descent method, where modification of the network branch weights corresponds to the negative gradient of the error. For weights on connections that directly connect to network output, this is straight forward. Error at the output layer is then propagated backwards to the hidden layer node.

According to the ANN shown in Fig. 3.1, the magnetization at any time instant $M(t)$ may be expressed in the form:

$$M(t) = f_{sig}\left(W_h \times f_{sig}\left(\sum_{\alpha_i \geq \beta_j} W_{i,j} \times P_{\alpha_i \beta_j} \right) \right) \qquad (3.5)$$

In the case when numerical values of the normalized training data are scaled down to be confined to the linear part of f_{sig}, (3.5) may be approximated as follows:

$$M(t) \approx \sum_{\alpha_i \geq \beta_j} \left[W_h \times W_{i,j} \times \left(\left. \frac{df_{sig}(x)}{dx} \right|_{x=0} \right)^2 \right] P_{\alpha_i \beta_j} \qquad (3.6)$$

Hence, the model unknown μ (α, β) for a pre-assumed $f(\varphi)$ may be directly computed from the ANN branch weights in accordance with the expression:

$$\mu(\alpha_i, \beta_j) \approx W_h \times W_{i,j} \times \left(\left. \frac{df_{sig}(x)}{dx} \right|_{x=0} \right)^2 \qquad (3.7)$$

It should be stated here that because the above-discussed training process is fast and fully automated, trying different pre-assumed $f(\varphi)$ functions should not impose significant computational burden in comparison to the more sophisticated identification activities associated with conventional vector hysteresis models.

To demonstrate the validity of the approach under consideration, numerical implementation and experimental testing were carried out for a hard disk sample [57]. More specifically, *M-H* curves measured at three different temperature values were used in the identification process. Normalized y-axis model input was, consequently, computed using the expression:

$$T = \left(\frac{Temperature - 25}{Temperature - 225} \right) \qquad (3.8)$$

The identification process was repeated using the above explained ANN training mechanism for different $f(\varphi)$ given by:

$$f(\varphi) = \cos^n(\varphi) \qquad (3.9)$$

where, n is an even integer.

Based upon the achieved mean squared error between the training measured and actual network outputs, it was found that $n = 12$ resulted in the best possible match. Sample computed *M-H* curves for different temperature values by the end of the identification process for the aforementioned n value are given in Fig. 3.2. Quantitative agreement of computed and measured curves have been demonstrated in [57].

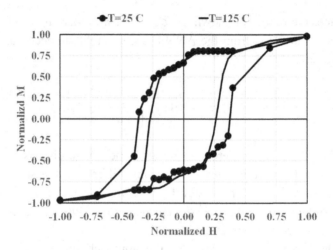

Figure 3.2. Sample computed M-H curves corresponding to the identification experimental data for different temperature values.

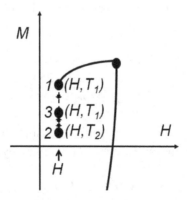

Figure 3.3. Typical irreversible M variation resulting from applying a T variation cycle while maintaining the applied input H value ($T_2 > T_1$).

The proposed model ability to assess M variations resulting arbitrary H-T changes was also checked. More specifically, simulations were carried out to check the model ability to predict model output M irreversible variations resulting from applying a T variation cycle corresponding to the y-axis input while maintaining the applied x-axis input H value. Sample results of this assessment are given in [57]. Those results are quantitatively similar to the typical variation shown in Fig. 3.3.

It can be concluded from the previously presented identification and simulation results as well as those reported in [57] that the proposed technique can lead to reasonably accurate qualitative and quantitative results. It should also be pointed out that the notion of using a two-dimensional model to simulate one-dimensional H-T-M variations may qualify a long list of unconventional simulation approaches, especially those reported in Chapter 2 of this book.

Chapter 4

Magnetic Hysteresis Models For Field-Mechanical Stress Variations

Magnetic materials exhibiting gigantic magnetostriction, especially Terfenol and Galfenol, are currently being frequently used in a wide range of applications. Examples of such applications include fine positioning actuators, sensors, energy harvesting, active vibration damping and assessment of optimum clamping stresses of dynamo and transformer core laminations. Obviously, using accurate magnetostriction simulation models during the design stages of devices utilized in the aforementioned applications is crucial for enhancing operational precision. In the past, numerous magnetostriction models having good qualitative agreement with experimentally observed data have been proposed (see, for instance, [7], [51], [58], [59], [60], [61], [62], [63], [64], [65], [66], [67] and [68]). In most of these models, however, accurate quantitative results can only be achieved through rigorous formulations and identification procedures that usually require the availability of specific measured data sets. In this Chapter some alternative unconventional models are presented where reasonable accuracy can be achieved while relaxing some of the identification data requirements and boosting computational efficiency. It should be mentioned here that the issue of computational efficiency becomes important when the ultimate goal is to incorporate a magnetostriction model in field computation tools.

4.1. A Feed-Forward Neural Network Magnetostriction Model

The concept of using two dimensional hysteresis models in simulating one dimensional magneto-elastic effects have been previously introduced in [58] and [59]. Based upon this concept, x- and y-axes model inputs are represented by the applied magnetic field H and the mechanical stress σ. Magnetization M and mechanical strain λ, on the other hand, represent the x- and y-axis model outputs. More precisely, magneto-elastic simulations may be carried out using the following representations:

$$M(t)$$

$$= \int_{-\pi/2}^{+\pi/2} \cos\varphi \left[\iint_{\alpha \geq \beta} \mu_1(\alpha,\beta)\, g_1(\varphi)\hat{\gamma}_{\alpha\beta} \left(\overline{e}_\varphi \cdot \left[H(t)\, \overline{e}_x + \sigma(t)\, \overline{e}_y \right] \right) d\alpha\, d\beta \right] d\varphi$$

(4.1)

$$\lambda(t)$$

$$= \int_{-\pi/2}^{+\pi/2} \sin\varphi \left[\iint_{\alpha \geq \beta} \mu_2(\alpha,\beta)\, g_2(\varphi)\hat{\gamma}_{\alpha\beta} \left(\overline{e}_\varphi \cdot \left[H(t)\, \overline{e}_x + \sigma(t)\, \overline{e}_y \right] \right) d\alpha\, d\beta \right] d\varphi$$

(4.2)

In these equations, t is the input-output time instant, $\hat{\gamma}_{\alpha\beta}$ are elementary rectangular hysteresis operators (as given in Fig. 1.3.a) having α and β switching up and down values, $\overline{e}_\varphi$ is a unit vector along the polar angle φ, g_1 and g_2 are pre-chosen symmetry functions, while μ_1 and μ_2 are the model unknowns that have to be determined through the identification procedure.

Interchanging the order of integration, we get:

$$M(t) = \iint_{\alpha \geq \beta} \mu_1(\alpha,\beta)\, I_{\alpha\beta}^M\, d\alpha\, d\beta$$

(4.3)

$$\lambda(t) = \iint_{\alpha \geq \beta} \mu_2(\alpha,\beta)\, I_{\alpha\beta}^\lambda\, d\alpha\, d\beta$$

(4.4)

where,

$$I_{\alpha\beta}^M = \int\limits_{-\pi/2}^{+\pi/2} \cos\varphi \, g_1(\varphi) \hat{\gamma}_{\alpha\beta} \left(\overline{e}_\varphi \cdot \left[H(t)\,\overline{e}_x + \sigma(t)\,\overline{e}_y \right] \right) d\varphi \qquad (4.5)$$

$$I_{\alpha\beta}^\lambda = \int\limits_{-\pi/2}^{+\pi/2} \sin\varphi \, g_2(\varphi) \hat{\gamma}_{\alpha\beta} \left(\overline{e}_\varphi \cdot \left[H(t)\,\overline{e}_x + \sigma(t)\,\overline{e}_y \right] \right) d\varphi \qquad (4.6)$$

Employing a finite number of elementary hysteresis operators, (4.3) – (4.6) may be reformulated as follows:

$$M(t) = \sum_{\alpha_i \geq \beta_j} \mu_1(\alpha_i, \beta_j) \, \mathrm{K}_{\alpha_i\beta_j}^M \qquad (4.7)$$

$$\lambda(t) = \sum_{\alpha_i \geq \beta_j} \mu_2(\alpha_i, \beta_j) \, \mathrm{K}_{\alpha_i\beta_j}^\lambda \qquad (4.8)$$

where,

$$\mathrm{K}_{\alpha_i\beta_j}^M = \int\limits_{-\pi/2}^{+\pi/2} \cos\varphi \, g_1(\varphi) \hat{\gamma}_{\alpha_i\beta_j} \left(\overline{e}_\varphi \cdot \left[H(t)\,\overline{e}_x + \sigma(t)\,\overline{e}_y \right] \right) d\varphi \qquad (4.9)$$

$$\mathrm{K}_{\alpha_i\beta_j}^\lambda = \int\limits_{-\pi/2}^{+\pi/2} \sin\varphi \, g_2(\varphi) \hat{\gamma}_{\alpha_i\beta_j} \left(\overline{e}_\varphi \cdot \left[H(t)\,\overline{e}_x + \sigma(t)\,\overline{e}_y \right] \right) d\varphi \qquad (4.10)$$

Given the coupled physical nature of this model, analytically solving its identification problem involves unquestionable complications. Moreover, availability of relevant experimental data specially tailored to accommodate the analytical solution of the aforementioned identification problem required sophisticated experimental setups.

It turns out that these difficulties may be overcome by utilizing artificial neural networks (ANN). In specific, the models given by (4.7) – (4.10) may be represented by the feed forward ANN (FFANN) shown in Fig. 4.1. This FFANN consists of an input stage, one hidden layer, and an output layer of neurons. Number of the input layer nodes is equivalent to the number of elementary hysteresis operators necessary to construct a relatively accurate model. On the other hand, each of the hidden and

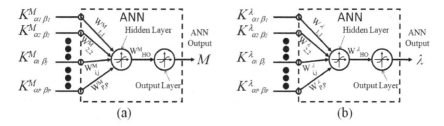

(a)   (b)

Figure 4.1. (a) FFANN correlating magnetization to applied magnetic field and mechanical stress, and (b) FFANN correlating mechanical strain to applied magnetic field and mechanical stress.

output layers contains a single neuron. The bipolar sigmoidal activation function f_{sig}, as given by (3.4), is employed throughout the FFANN.

In the presence of a set of measured first order reversal curves under different mechanical stress values of the orthogonal model parameter corresponding to the model used, and by assuming specific symmetry functions g_1 and g_2, integration values $K^M_{\alpha_i \beta_j}$ and $K^\lambda_{\alpha_i \beta_j}$ may be computed at every time instant t. Thus, two sets of training (identification) input-output data required by the FFANNs to determine their optimum branch weights are generated. This training is performed using a typical ANN back-propagation supervised learning algorithm [47]. Starting from some random branch weight values, training repeatedly modifies these weights. In any one of the two FFANNs, the feed-forward process of the training presents an input pattern to input layer nodes that pass the input values onto the hidden layer where a weighted summation operation is performed through the activation function.

Likewise, the network output is computed through the activation function of the output layer node. Since the ultimate identification goal is to make the FFANN output as close as possible to the measured identification data, the cumulative mean squared error between the network training and computed outputs is minimized based on the gradient descent method. In other words, network weights are modified during the training process in a direction that corresponds to the negative gradient of the error (see, for example, [32], [33] and [47]). Upon achieving an acceptable cumulative mean squared error, the identification process is terminated. Although simulation may then be directly carried out using the model in its FFANN form, it is possible to map the deduced branch weights to the model unknowns. More specifically, the model unknowns may be deduced as follows:

$$M(t) = f_{sig}\left(W_{HO}^M \times f_{sig}\left(\sum_{\alpha_i \geq \beta_j} W_{\alpha_i,\beta_j}^M \times K_{\alpha_i \beta_j}^M \right) \right) \qquad (4.11)$$

$$\lambda(t) = f_{sig}\left(W_{HO}^\lambda \times f_{sig}\left(\sum_{\alpha_i \geq \beta_j} W_{\alpha_i,\beta_j}^\lambda \times K_{\alpha_i \beta_j}^\lambda \right) \right) \qquad (4.12)$$

In the case when the training data is intentionally scaled down, the previous expressions may be approximated as:

$$M(t) \approx \sum_{\alpha_i \geq \beta_j}\left\{ W_{\alpha_i,\beta_j}^M \times W_{HO}^M \times \left(\frac{df_{sig}(x)}{dx}\bigg|_{x=0} \right)^2 \right\} K_{\alpha_i \beta_j}^M \qquad (4.13)$$

$$\lambda(t) \approx \sum_{\alpha_i \geq \beta_j}\left\{ W_{\alpha_i,\beta_j}^\lambda \times W_{HO}^\lambda \times \left(\frac{df_{sig}(x)}{dx}\bigg|_{x=0} \right)^2 \right\} K_{\alpha_i \beta_j}^\lambda \qquad (4.14)$$

Thus,

$$\mu_1(\alpha_i, \beta_j) \approx W_{\alpha_i,\beta_j}^M \times W_{HO}^M \times \left(\frac{df_{sig}(x)}{dx}\bigg|_{x=0} \right) \qquad (4.15)$$

$$\mu_2(\alpha_i, \beta_j) \approx W_{\alpha_i,\beta_j}^\lambda \times W_{HO}^\lambda \times \left(\frac{df_{sig}(x)}{dx}\bigg|_{x=0} \right) \qquad (4.16)$$

In order to assess the accuracy and practicality of the model under consideration, numerical implementation and experimental testing were performed as reported in [57]. Measured *M-H* and *λ-H* curves for a Terfenol-D rod, under different stress values ranging from 0.9347 to 34.512 Kpsi were used for the identification. Sample simulation results corresponding to the identification (training) experimental data for the pre-stress and maximum stress cases are shown in Fig. 4.2. These results were computed for the pre-selected symmetry functions given by $g_1(\varphi) = \cos^2(\varphi)$ and $g_2(\varphi) = \cos(\varphi)$. An important observation is the fact that a negative y-axis input at zero applied x-axis input naturally results in a negative y-axis output. Applying a cyclic x-axis input clearly results in the tracing the expected and well-known butterfly-type y-axis output which is in full qualitative harmony with experimentally observed magnetostriction curves as previously shown in Fig. 1.8. More testing results may be found in [57].

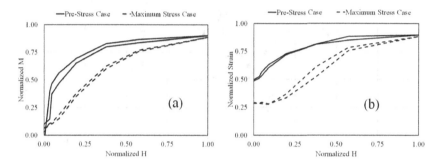

Figure 4.2. Sample simulation results corresponding to the identification (training) experimental data for the pre-stress and maximum stress cases; (a) *M-H* curves and, (b) *λ-H* curves.

4.2. Implementation of Magnetostriction Models Using Orthogonally Coupled Hysteresis Operators

It has been shown in Section 1.5 that a single elementary hysteresis operator may be realized via a dual-node discrete Hopfield Neural Network (DHNN). Let us now consider the four-node DHNN network shown in Fig. 4.3. Based upon the proposed approaches discussed in [41] and [58], the DHNN shown in this figure may be employed to qualitatively construct a magnetostriction model whose inputs I_x and I_y correspond to the magnetic field H and mechanical stress σ, while its outputs denoted by O_x and O_y represent the magnetization M or mechanical strain λ. For the DHNN under consideration, $k_{//}$ and $k_{\perp}$ represent feedback factors corresponding to similar and dissimilar physical quantities, respectively.

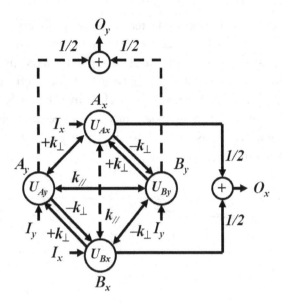

Figure 4.3. The four-node DHNN representing the orthogonally coupled hysteresis operators.

In accordance with the basic configuration of a typical hysteresis model, an ensemble of elementary operators having different switching up and down thresholds is usually required (refer, for instance, to [6]). This may be achieved in the proposed model by tuning the operator widths controlled by $k_{//}$ and applying offset terms OS to the model inputs. Overall, the suggested model may be implemented using the modular combination of DHNN blocks through two linear Neural Network (LNN) stages, thus forming the DHNN-LNN configuration shown in Fig. 4.4. In this figure, offset values OS_i and feedback factors $k_{//i}$ corresponding to the particular i^{th} DHNN block corresponding to the rectangular operator whose switching up and down thresholds are α_i and β_i, respectively, may be given by:

$$OS_i = -\frac{(\alpha_i + \beta_i)}{2} \tag{4.17}$$

$$k_{//i} = +\frac{(\alpha_i - \beta_i)}{2} \tag{4.18}$$

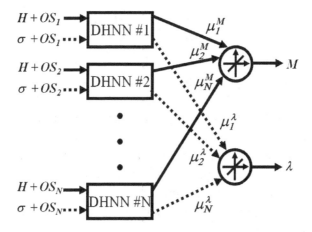

Figure 4.4. Realization of the proposed model using a DHNN-LNN configuration.

For any applied inputs, the DHNN blocks outputs, comprising a finite set of N operators, evolve such that a minimum state is achieved for the overall network quadratic energy E given by:

$$E = -\sum_{i=1}^{N} \left\{ \begin{array}{c} +\left(H - \left[\dfrac{\alpha_i + \beta_i}{2}\right]\right)\left(U_{Axi} + U_{Bxi}\right) + \left[\dfrac{\alpha_i - \beta_i}{2}\right]U_{Axi}U_{Bxi} \\[2.5ex] +\left(\sigma - \left[\dfrac{\alpha_i + \beta_i}{2}\right]\right)\left(U_{Ayi} + U_{Byi}\right) + \left[\dfrac{\alpha_i - \beta_i}{2}\right]U_{Ayi}U_{Byi} \\[2.5ex] +\dfrac{k_\perp}{2}\left(U_{Axi} - U_{Bxi}\right)\left(U_{Ayi} + U_{Byi}\right) \\[2.5ex] +\dfrac{k_\perp}{2}\left(U_{Ayi} - U_{Byi}\right)\left(U_{Axi} + U_{Bxi}\right) \end{array} \right\} \qquad (4.19)$$

where, H is the applied magnetic field and the σ is the applied mechanical stress.

For the DHNN-LNN configuration, the model outputs may be expressed in the form:

$$M = \sum_{i=1}^{N} \mu_i^M \left(\frac{U_{Axi} + U_{Bxi}}{2}\right) \qquad (4.20)$$

$$\lambda = \sum_{i=1}^{N} \mu_i^\lambda \left(\frac{U_{Ayi} + U_{Byi}}{2} \right) \qquad (4.21)$$

In (4.20) and (4.21) μ_i^M and μ_i^λ, together with $k_\perp$, represent the model unknowns that have to be determined through the identification process such a best fit to the experimental data of any material under consideration is achieved.

Without resorting to cumbersome analytical approaches, this identification may be easily achieved for any pre-assumed $k_\perp$ value by employing standard NN training procedures.

Numerical implementation example of the proposed modeling strategy was carried out for a Terfenol material rod. In this implementation, about 900 uniformly dispersed elementary operators – having different offset magnitudes and loop widths controlled by the various $k_{//i}$ values – were utilized. The model unknowns were identified for different pre-chosen $k_\perp$ values. More specifically, the identification process was carried out through a typical training process for the previously discussed DHNN-LNN configurations [33]. In this training process, two sets of *M-H* and *λ-H* curves, each corresponding to two different constant mechanical stress values, were utilized. Hence, potential complications associated with the analytical solution of the identification problem for such a model configuration is bypassed. Another advantage of this DHNN-LNN approach is the possibility of carrying out the identification process using any available set of experimental data. Obviously, the more representative this set is the more accurate the model will be to simulate other input-output variations. Example of the identification process outcome is shown in Fig. 4.5. It should be stated that the applied magnetic field and mechanical stress values in this figure were normalized to 1.38 KN/cm² and 3000 Oe, respectively. It should also be mentioned that the quantitative agreement shown in this figure for this particular material may be, more or less, achieved for $k_\perp$ values ranging between 1.0 and 8.0.

In order to assess the optimum $k_\perp$ value for the material under consideration, comparison between computed and additional identification measured data not involved in the training process should be employed. For the implementation case under consideration, it was found that $k_\perp \approx 5.0$ would lead to the best possible quantitative match for *M-H* and λ-*H* curves corresponding to a constant normalized mechanical stress of 0.5 as demonstrated in Fig. 4.6. Similar implementation results for a Terfenol sample have also been reported in [69].

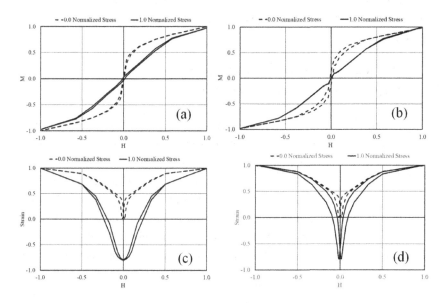

Figure 4.5. Sample results highlighting the outcome of the DHNN-LNN training process for two different applied stress values; (a) measured *M-H* curves, (b) computed *M-H* curves, (c) measured λ-*H* curves, and (d) computed λ-*H* curves.

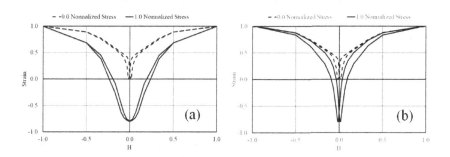

Figure 4.6. Comparison between (a) measured and (b) computed results for an applied mechanical stress not involved in the DHNN-LNN training process.

A slight variation of the model under consideration was reported in [70] where discrete HNNs where replaced by the hybrid activation functions leading to Stoner-Wohlfarth-like elementary operators as previously discussed in Section 1.6. Given the smooth nature of such operators, it is possible to significantly reduce the number of elementary operators required to construct a hysteresis curve since no operators would be needed for loop edges where input-output relations are single valued.

4.3. Magnetostriction Modelling Using an Ensemble of Magnetized Triangular Subregions

It has been shown in Section 2.5 that 2D magnetized geometrical configurations may be realized by an ensemble of tri-node HNNs each representing a triangular subregion as demonstrated in Figs. 2.28 and 2.29. It turns out that the same approach has been employed to present an efficient magneto-elastic material simulation methodology [71]. More specifically, adopting the concept proposed in [58] – where 1D magnetostriction is simulated by using a 2D vector hysteresis model – a tri-node HNN input $\bar{h}$ and overall output $\bar{m}$ may be represented by the expression:

$$\begin{bmatrix} \bar{h} \\ \bar{m} \end{bmatrix} = \begin{bmatrix} H\,\bar{e}_x + \sigma\,\bar{e}_y \\ M\,\bar{e}_x + \lambda\,\bar{e}_y \end{bmatrix} \tag{4.22}$$

where, H is the applied magnetic field, σ is the applied mechanical stress, M is the magnetization, λ is the mechanical strain, while $\bar{e}_x$ and $\bar{e}_y$ represent the unit vectors along the x- and y-directions, respectively. As previously discussed, for any applied input the tri-node HNN converges to the overall output corresponding to an overall minimum energy E given by:

$$E = -\sum_{i=1}^{3}\left\{ \bar{h} \bullet \bar{m}_i + \frac{1}{2}\sum_{j=1, j\neq i}^{3} k_{i,j}\, m_i m_j \right\} \tag{4.23}$$

In (4.23) at any time instant t, for chosen values of inter-node coupling coefficient k_{copl} and a per-unit value c of the continuous node activation function we have:

$$k_{i,j} = k_{copl} \left| \cos(\varphi_{ij}) \right| \qquad (4.24)$$

$$m_i(t+1) = c\, f_c(net_i(t)) + (1-c)\, f_d(net_i(t)) \qquad (4.25)$$

$$f_d(net_i(t)) = \mathrm{sgn}(net_i(t)) = \begin{cases} +1 & \text{if } net_i(t) > 0 \\ -1 & \text{if } net_i(t) < 0 \\ \text{Unchanged} & \text{if } net_i(t) = 0 \end{cases} \qquad (4.26)$$

$$f_c(net_i(t)) = \tanh\left(a[net_i(t)] \right) \qquad (4.27)$$

$$net_i(t) = \bar{h}(t) \bullet \left(\frac{\bar{m}_i}{m_i} \right) + \sum_{j=1, j \neq i}^{3} k_{i,j}\, m_i(t) \qquad (4.28)$$

The overall output of the tri-node HNN given in (4.22) may be computed from the expression:

$$\bar{m} = \sum_{i=1}^{3} \bar{m}_i \qquad (4.29)$$

In order to demonstrate the qualitative nature of this approach, two different tri-node HNN configurations – one corresponding to an equilateral triangle and the other corresponding to an isosceles triangle – are considered. More specifically, angular orientations of the nodes of the two aforementioned HNNs are assumed to be given by 00.0, +120.0, -120.0 and 00.0, +100.0, -100.0, respectively. Traced *M-H* and *λ-H* curves for both tri-node HNNs having different parameters and corresponding to constant normalized mechanical stress values of 0.2, 0.4 and 0.6 are shown in Figs. 4.7 and 4.8. It is clear from these figures that the model parameters may be tuned to yield different *M-H* and *λ-H* curve profiles. The figures also demonstrate how the triangular shape configuration affects the aforementioned profiles.

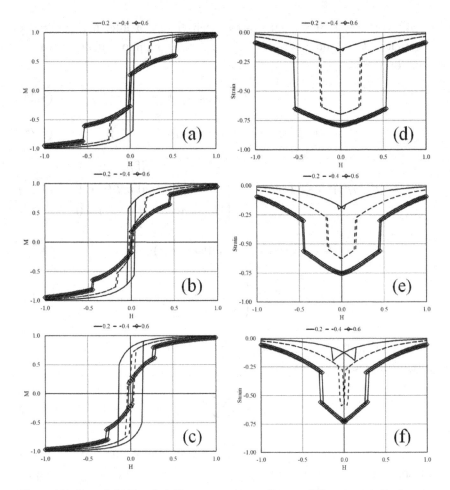

Figure 4.7. Traced *M-H* and *λ-H* curves corresponding to different normalized stress values for the tri-node HNN having node angular orientations given by 00.0, +120.0, -120.0; (a) & (d) k_{copl} = 0.25, c = 0.5, a = 3.0, (b) & (e) k_{copl} = 0.3, c = 0.7, a = 3.0, (c) & (f) k_{copl} = 0.4, c = 0.7, a = 3.0.

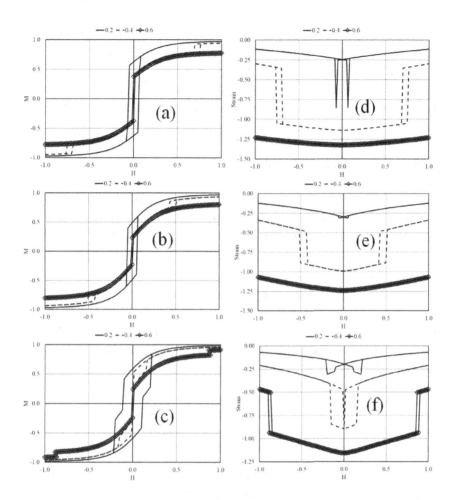

Figure 4.8. Traced M-H and λ-H curves corresponding to different normalized stress values for the tri-node HNN having node angular orientations given by 00.0, +100.0, -100.0; (a) & (d) k_{copl} = 0.25, c = 0.5, a = 3.0, (b) & (e) k_{copl} = 0.3, c = 0.7, a = 3.0, (c) & (f) k_{copl} = 0.4, c = 0.7, a = 3.0.

It turns out that by using an ensemble of the previously discussed triangular shaped tri-node HNNs in the form of an ellipse, a more generalized magnetostriction model may be obtained. Moreover, changing the aspect ratio of the ellipse axes will affect the various triangular angular configurations, the fact that will result in adding an additional tuning parameter to fit measurements corresponding to a particular material. In order to demonstrate the ability of the proposed approach to simulate different magneto-elastic materials, subject to both magnetic field and mechanical stress variations, numerous computations were performed while varying the various parameters to highlight the possibility of tuning the model quantitative results. Throughout these simulations, λ-H variations at two pre-chosen stress values of 0.15 and 0.30 were normalized to the corresponding strain amplitude for an assumed 0.02 pre-stress value.

Simulations were carried for model parameter a, c and k_{copl} values given by: $a = 3.0, 9.0 : k_{copl} = 0.02, 0.04, 0.06 : c = 0.1, 0.3, 0.5$. Moreover, an elliptical ensemble of 36 triangular shaped tri-node HNNs was chosen to have x-axis to y-axis diameter aspect ratios of 1:1, 4:1 and 1:4 (see Fig. 2.36.a). In this case, total model output may be given by the expression:

$$\bar{M} = \sum_{j=1}^{36} \bar{m}_j = \sum_{j=1}^{36} \left(\sum_{i=1}^{3} \bar{m}_{j,i} \right) \tag{4.30}$$

where, $\bar{m}_j$ and $\bar{m}_{j,i}$ represent the total output of the triangular tri-node HNN number j and its corresponding node outputs, respectively.

Sample simulation results are shown in Figs. 4.9 – 4.20. These results clearly reflect the possibility of mimicking different M-H and λ-H variations through model parameter variations. It should be mentioned here that a 3D extension of the concept of using a triangular shaped tri-node HNN was recently proposed and employed in magnetostriction modeling. Within this extension, the applied stress effect was assumed to shift a tripod centroid vertically thus changing the final orientations of the various activation functions and, consequently, intercoupling factors. Results of this extension are, more or less, similar to those presented in this Section (see [72]).

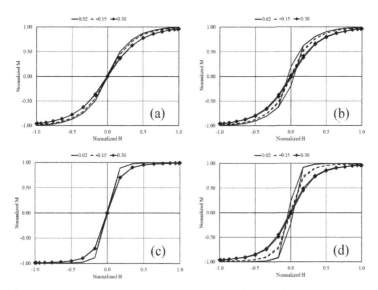

Figure 4.9. Traced *M-H* curves corresponding to an elliptical ensemble of 36 triangular tri-node HNNs having 1:1 diameter aspect ratio and k_{copl} = 0.02; (a) c = 0.1, a = 3.0, (b) c = 0.3, a = 3.0, (c) c = 0.1, a = 9.0, and (d) c = 0.3, a = 9.0.

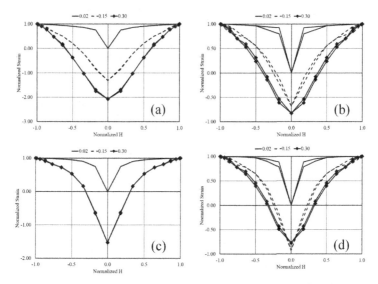

Figure 4.10. Traced *λ-H* curves corresponding to an elliptical ensemble of 36 triangular tri-node HNNs having 1:1 diameter aspect ratio and k_{copl} = 0.02; (a) c = 0.1, a = 3.0, (b) c = 0.3, a = 3.0, (c) c = 0.1, a = 9.0, and (d) c = 0.3, a = 9.0.

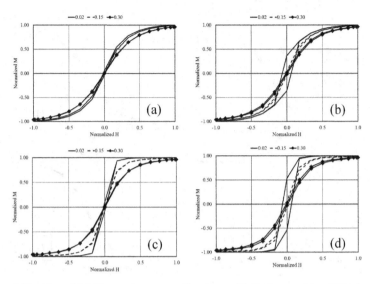

Figure 4.11. Traced *M-H* curves corresponding to an elliptical ensemble of 36 triangular tri-node HNNs having 1:1 diameter aspect ratio and k_{copl} = 0.04; (a) c = 0.1, a = 3.0, (b) c = 0.3, a = 3.0, (c) c = 0.1, a = 9.0, and (d) c = 0.3, a = 9.0.

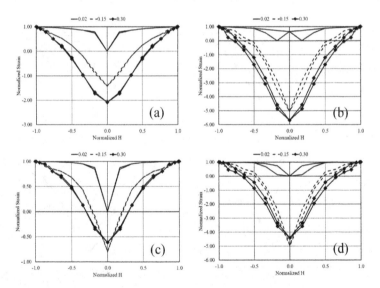

Figure 4.12. Traced *λ-H* curves corresponding to an elliptical ensemble of 36 triangular tri-node HNNs having 1:1 diameter aspect ratio and k_{copl} = 0.04; (a) c = 0.1, a = 3.0, (b) c = 0.3, a = 3.0, (c) c = 0.1, a = 9.0, and (d) c = 0.3, a = 9.0.

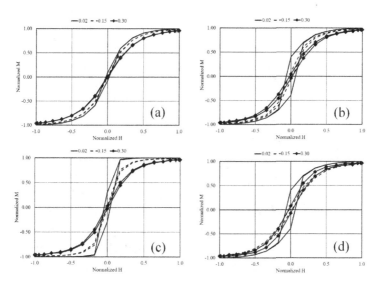

Figure 4.13. Traced *M-H* curves corresponding to an elliptical ensemble of 36 triangular tri-node HNNs having 1:1 diameter aspect ratio and k_{copl} = 0.06; (a) c = 0.1, a = 3.0, (b) c = 0.3, a = 3.0, (c) c = 0.1, a = 9.0, and (d) c = 0.3, a = 9.0.

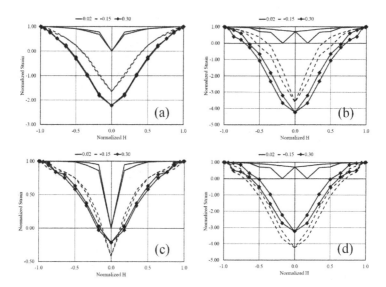

Figure 4.14. Traced *λ-H* curves corresponding to an elliptical ensemble of 36 triangular tri-node HNNs having 1:1 diameter aspect ratio and k_{copl} = 0.06; (a) c = 0.1, a = 3.0, (b) c = 0.3, a = 3.0, (c) c = 0.1, a = 9.0, and (d) c = 0.3, a = 9.0.

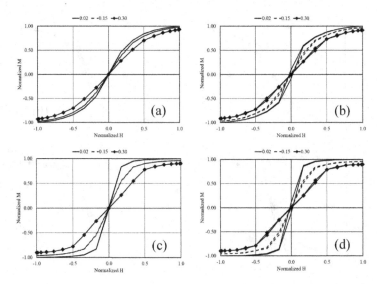

Figure 4.15. Traced *M-H* curves corresponding to an elliptical ensemble of 36 triangular tri-node HNNs having 1:4 diameter aspect ratio and k_{copl} = 0.02; (a) c = 0.1, a = 3.0, (b) c = 0.3, a = 3.0, (c) c = 0.1, a = 9.0, and (d) c = 0.3, a = 9.0.

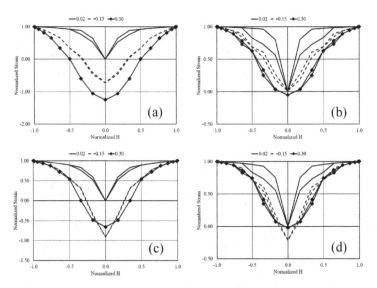

Figure 4.16. Traced λ-*H* curves corresponding to an elliptical ensemble of 36 triangular tri-node HNNs having 1:4 diameter aspect ratio and k_{copl} = 0.02; (a) c = 0.1, a = 3.0, (b) c = 0.3, a = 3.0, (c) c = 0.1, a = 9.0, and (d) c = 0.3, a = 9.0.

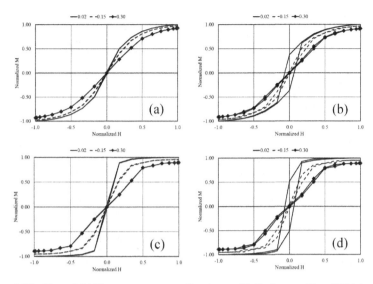

Figure 4.17. Traced *M-H* curves corresponding to an elliptical ensemble of 36 triangular tri-node HNNs having 1:4 diameter aspect ratio and k_{copl} = 0.04; (a) c = 0.1, a = 3.0, (b) c = 0.3, a = 3.0, (c) c = 0.1, a = 9.0, and (d) c = 0.3, a = 9.0.

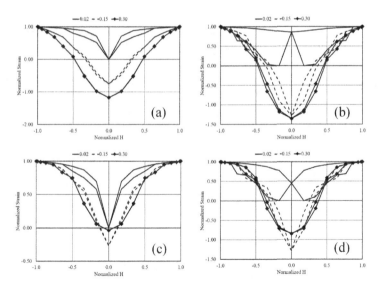

Figure 4.18. Traced *λ-H* curves corresponding to an elliptical ensemble of 36 triangular tri-node HNNs having 1:4 diameter aspect ratio and k_{copl} = 0.04; (a) c = 0.1, a = 3.0, (b) c = 0.3, a = 3.0, (c) c = 0.1, a = 9.0, and (d) c = 0.3, a = 9.0.

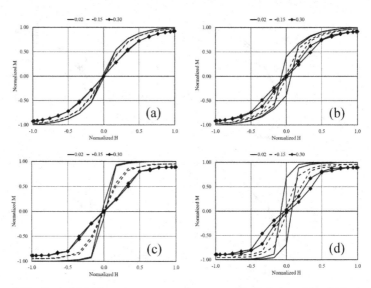

Figure 4.19. Traced *M-H* curves corresponding to an elliptical ensemble of 36 triangular tri-node HNNs having 1:4 diameter aspect ratio and k_{copl} = 0.06; (a) c = 0.1, a = 3.0, (b) c = 0.3, a = 3.0, (c) c = 0.1, a = 9.0, and (d) c = 0.3, a = 9.0.

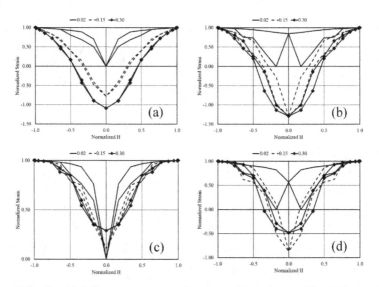

Figure 4.20. Traced *λ-H* curves corresponding to an elliptical ensemble of 36 triangular tri-node HNNs having 1:4 diameter aspect ratio and k_{copl} = 0.06; (a) c = 0.1, a = 3.0, (b) c = 0.3, a = 3.0, (c) c = 0.1, a = 9.0, and (d) c = 0.3, a = 9.0.

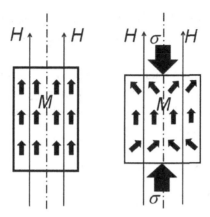

Figure 4.21. The expected domain rotation resulting from applying a mechanical stress σ to an axially magnetized rod subject to an applied magnetic field H.

4.4. A Domain-Rotation-Inspired Magnetostriction Model

It is well known that applying a mechanical stress along the axial direction of a magnetized rod results in a magnetization reduction as a result of off-axis domain rotation as shown in Fig. 4.21 (refer, for instance, to [20]). As can be seen in this figure, diagonally opposing domains experience mechanical stress-induced rotation in the same direction. Stemming from this fact another approach to model magnetostriction was recently proposed in [73]. In this approach a pair of identical 2D vector hysteresis models are employed, each representing diagonally combined zones of the material under consideration. For the model representing the zones expected to experience domain rotation in the clockwise direction, x and y inputs correspond to the applied mechanical stress and applied magnetic field respectively. On the other hand, x and y inputs of the model representing the zones expected to experience domain rotation in the counterclockwise direction correspond to the negative value of applied mechanical stress and applied magnetic field respectively. Inputs of both models may, hence, be given by:

$$\bar{u}_A = +\sigma\,\bar{u}_x + H\,\bar{u}_y \tag{4.31}$$

$$\bar{u}_B = -\sigma\,\bar{u}_x + H\,\bar{u}_y \tag{4.32}$$

where suffixes A and B denote the zones experiencing clockwise and counter clockwise domain rotation as a result of applying a compressive stress, respectively.

Employing the previously proposed model comprised of a rotationally oriented group of primitive vector hysteresis models, each realized using a tri-node HNN, details of the model corresponding to zone A may be given by the expressions:

$$\overline{f}_A = \sum_{s=1}^{N} \left| \overline{m}_s(\overline{u}_A) \right| e^{j\theta_s} = \sum_{s=1}^{N} \left| \sum_{t=1}^{3} m_{st}(\overline{u}_A)\overline{e}_{st} \right| e^{j\theta_s} \qquad (4.33)$$

$$m_{st} = c\zeta_c(\omega_{st}) + (1-c)\zeta_d(\omega_{st}), \quad \theta_s = e^{j(s-1)\frac{2\pi}{N}} \qquad (4.34)$$

$$\zeta_c(\omega) = \tanh(a\omega), \quad a > 0 \qquad (4.44)$$

$$\zeta_d(\omega) = \begin{cases} +1 & \text{if } \omega > 0 \\ -1 & \text{if } \omega < 0 \\ \text{unchanged} & \text{if } \omega = 0 \end{cases} \qquad (4.45)$$

$$\omega_{st} = \overline{u}_A \bullet \overline{e}_{st} + \sum_{r=1,r\neq t}^{3} m_{sr}k_{copl} \left| \overline{e}_{sr} \bullet \overline{e}_{st} \right| \qquad (4.46)$$

where, m, ω, ζ_c and ζ_d are the HNN node output, input, continuous activation function, and discrete activation function, respectively, $\overline{e}_{st}$ is the unit vector identifying the HNN node t output orientation within the tri-node HNN s, while a, c, and k_{copl} are the model unknowns that need to be determined through the identification process [71].

Based upon this implementation, each tri-node HNN converges to the state of minimum energy E_s given by:

$$E_s = -\sum_{t=1}^{3} \left[\overline{u}_A \bullet m_{st}\overline{e}_{st} + \frac{1}{2} \sum_{r=1,r\neq t}^{3} m_{sr}m_{st}k_{copl} \left| \overline{e}_{sr} \bullet \overline{e}_{st} \right| \right] \qquad (4.47)$$

Likewise, similar expressions may be inferred for the model corresponding to zone B where its output may be expressed in the form:

$$\overline{f}_B = \sum_{s=1}^{N} \left| \overline{m}_s(\overline{u}_B) \right| e^{j\theta_s} = \sum_{s=1}^{N} \left| \sum_{t=1}^{3} m_{st}(\overline{u}_B)\overline{e}_{st} \right| e^{j\theta_s} \qquad (4.48)$$

Qualitatively speaking, the overall model output computed using the superposition principle may be given by:

$$\begin{bmatrix} M_q(\sigma, H) \\ \lambda_q(\sigma, H) \end{bmatrix} = \begin{bmatrix} \dfrac{\overline{f}_A(\overline{u}_A) + \overline{f}_B(\overline{u}_B)}{2} \\ \dfrac{\overline{f}_A(\overline{u}_A) - \overline{f}_B(\overline{u}_B)}{2} \end{bmatrix} \qquad (4.49)$$

where, M_q and λ_q represent the qualitative model outputs corresponding to the magnetization and mechanical strain, respectively.

There is no doubt that by tuning the model parameters, quantitative match of *M-H* and *λ-H* curves with experimental data at a particular applied mechanical stress may be easily achieved. While qualitative match for similar identification curves corresponding to other applied mechanical stress values is guaranteed by using the model under consideration, quantitative match may be significantly enhanced through the utilization of scaling polynomials. Since these polynomials are single valued functions, they basically act on the anhysteretic components of the curves. Denoting the magnetization and mechanical strain scaling polynomials by P^M and P^λ, respectively, a possible expression for the model output may expressed in the form:

$$M(\sigma, H) = \sum_{i=0}^{N_\sigma} \sum_{j=0}^{N_M} P_{i,j}^M \sigma^i \left[M_q(\sigma, H) \right]^j \qquad (4.50)$$

$$\lambda(\sigma, H) = \sum_{i=0}^{N_\sigma} \sum_{j=0}^{N_\lambda} P_{i,j}^\lambda \sigma^i \left[\lambda_q(\sigma, H) \right]^j \qquad (4.51)$$

where, N_σ, N_M, and N_λ are the polynomial degrees, while $P_{i,j}^M$ and $P_{i,j}^\lambda$ represent the various fitting polynomial coefficients.

In the case when the minimum applied mechanical stress value (i.e., pre-stress value for instance) for the identification *M-H* and *λ-H* curves is denoted by $\sigma_{\min}$ and knowing that the main goal is to quantitatively tune the model output corresponding to other applied stress values, a possible form of $M(\sigma, H)$ may be given by:

$$M(\sigma,H) = M_q(\sigma,H) - \left\{ C\left[\frac{\sigma - \sigma_{\min}}{\sigma_{\min}}\right] M_q(\sigma,H)\left(1 - \left[M_q(\sigma,H)\right]^{l-1}\right)\right\}$$

(4.52)

where, l is an odd integer and C is some scaling constant.

In addition to the aforementioned goal of maintaining *M-H* curve corresponding to the minimum applied stress value unscaled, it maintains the odd nature of the *M-H* curves and offers minimal scaling intervention near the maximum applied field amplitude. In the case when expression (4.52) is adopted, the various polynomial coefficients $P_{i,j}^M$ may thus be given by:

$$P_{i,j}^M = \begin{cases} (1+C) & i=0, j=1 \\[2mm] -\dfrac{C}{\sigma_{\min}} & i=1, j=1 \\[2mm] -C & i=0, j=l \\[2mm] \dfrac{C}{\sigma_{\min}} & i=1, j=l \\[2mm] 0 & \text{Otherwise} \end{cases}$$

(4.53)

In order to demonstrate the proposed approach, simulations were carried out to mimic the *M-H* curves of a typical Terfenol sample corresponding to normalized stress values of 0.1, 0.2 and 0.3. Sample simulation results corresponding to these stress values for different values of l and C are shown in Figs. 4.22 and 4.23. From the demonstrated results, it is clear that quantitative nature of the curves corresponding to stress values other than 0.1 may be tuned. Obviously using more sophisticated polynomial scaling functions could offer additional flexibility in tuning each curve for different applied field intervals.

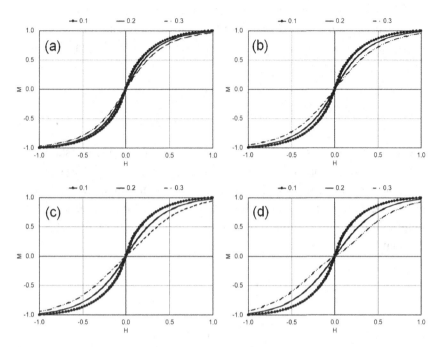

Figure 4.22. Sample M-H simulation results corresponding to different normalized stress values for, (a) No scaling applied, (b) C = 0.1 and l = 3, (c) C = 0.2 and l = 3, (c) C = 0.3 and l = 3.

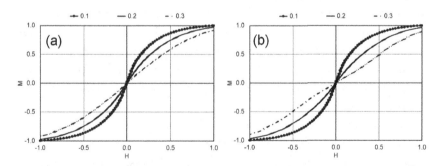

Figure 4.23. Sample M-H simulation results corresponding to different normalized stress values for, (a) C = 0.2 and l = 5, (b) C = 0.3 and l = 5.

Likewise, knowing that the main goal is to quantitatively tune computed λ-H curves corresponding to identification data for mechanical stress values other than $\sigma_{\min}$, a possible form of $\lambda(\sigma, H)$ may be given by:

$$\lambda(\sigma, H) = \lambda_q(\sigma, H) + C\left[\frac{\sigma - \sigma_{\min}}{\sigma_{\min}}\right]\left[\lambda_q(\sigma, H)\right]^l \qquad (4.54)$$

where, once more, l is an odd integer and C is some scaling constant.

This formulation would result in maintaining λ-H curve corresponding to the minimum applied stress value unscaled in addition to preserving the even nature of the λ-H curves. For the particular representation given by expression (4.54), the various polynomial coefficients $P_{i,j}^{\lambda}$ may thus be given by:

$$P_{i,j}^{\lambda} = \begin{cases} 1 & i = 0, j = 1 \\ -C & i = 0, j = l \\ \dfrac{C}{\sigma_{\min}} & i = 1, j = l \\ 0 & \text{Otherwise} \end{cases} \qquad (4.55)$$

Sample typical Terfenol λ-H simulation results corresponding to normalized stress values of 0.1, 0.2 and 0.3 for different values of l and C are shown in Figs. 4.24 and 4.25. It can be seen again from these results that curves corresponding to normalized stress values other than 0.1 may be quantitatively tuned. More sophisticated scaling polynomials may definitely be utilized to fine tune λ-H at different applied field values.

It should be mentioned that an effort to assess the validity and accuracy of proposed magnetostriction model had been reported in [73]. It can be inferred from the presented model details, simulation results, and comparisons with measured curves results reported in [73] that the proposed hysteresis modeling approach may be regarded as an efficient and considerably accurate tool to simulate unidirectional magneto-elastic interactions.

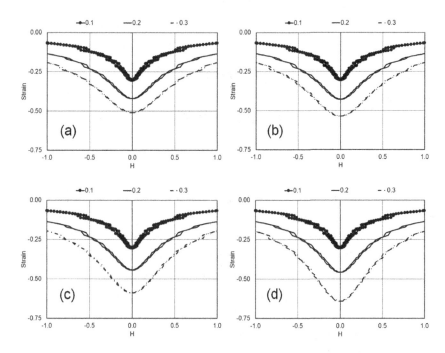

Figure 4.24. Sample λ-H simulation results corresponding to different normalized stress values for, (a) No scaling applied, (b) C = 0.1 and l = 3, (c) C = 0.3 and l = 3, (c) C = 0.5 and l = 3.

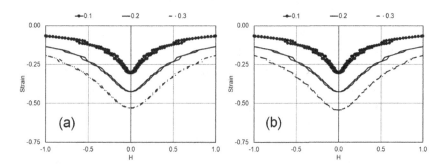

Figure 4.25. Sample M-H simulation results corresponding to different normalized stress values for, (a) C = 0.3 and l = 5, (b) C = 0.5 and l = 5.

Chapter 5

Conclusions

In the previous chapters of this book some efficient and unconventional models to simulate multi-component magnetic hysteresis have been discussed and presented. Several points, however, have to be stressed to clarify the potential utilization applicability and justification of those models. These points may be summarized as follows:

- The proposed models may not be regarded, in any way, as a substitute to more rigorous mathematically formulated phenomenological models (such as [7], [30] and [74]) whenever their necessary identification physical and/or experimental data is available.

- In the presence of partial and/or nonsystematic experimental identification data that prohibit usage of rigorous phenomenological models, the various modeling approaches presented in this book may be alternatively utilized.

- Since the identification processes of the proposed models are, in general, carried out through training processes that utilize any available experimental data it should be understood that the expected simulation accuracy would be dependent on the extent of how the utilized identification/training data represent the various material magnetic properties.

- Since the proposed models are partially (or sometimes fully) realized using artificial neural networks, it is possible to utilize a number of relevant commercially available software packages (that include neural network modules) in constructing, identifying, and running these models.

- Whenever the final goal is to integrate hysteresis models into field computation packages for device design and/or overall performance assessment, computational efficiency of the proposed models in exchange for an acceptable less accuracy margin may be regarded as an advantage over other rigorous models that require relatively high computational resources.

While some out of the box unconventional ideas to model multi-component hysteresis have been assembled in this book, there is no doubt that the efforts will continue to achieve higher accuracies at less computational costs. Finally, it should be mentioned that the choice of the most suitable multi-component hysteresis model is, eventually, dependent on the user technical background in addition to the invariant factors related to the specific application and availability of data.

Bibliography

[1] E. Madelung, "On the magnetization by a fast currents and an operation principle of the magnetodetectors of Rutherford-Marconi,' (in German)," *Ann. Phys.*, vol. 17, no. 5, pp. 861–890, 1905.

[2] B. C. O'Regan, P. R. F. Barnes, X. Li, C. Law, E. Palomares, and J. M. Marin-Beloqui, "Optoelectronic studies of methylammonium lead iodide perovskite solar cells with mesoporous TiO2: Separation of electronic and chemical charge storage, understanding two recombination lifetimes, and the evolution of band offsets during J - V hysteresis," *J. Am. Chem. Soc.*, 2015, doi: 10.1021/jacs.5b00761.

[3] F. Al-Bender, W. Symens, J. Swevers, and H. Van Brussel, "Theoretical analysis of the dynamic behavior of hysteresis elements in mechanical systems," *Int. J. Non. Linear. Mech.*, 2004, doi: 10.1016/j.ijnonlinmec.2004.04.005.

[4] I. D. Mayergoyz, "Superconducting hysteresis," *The Science of Hysteresis*, 2006.

[5] E. C. Stoner and E. P. Wohlfarth, "A mechanism of magnetic hysteresis in heterogeneous alloys," *Philos. Trans. R. Soc. A*, vol. 240, no. 826, pp. 599–642, 1948.

[6] I. D. Mayergoyz, *Mathematical Models of Hysteresis and Their Applications*. 2003.

[7] I. D. Mayergoyz, "Mathematical models of hysteresis (Invited)," *IEEE Trans. Magn.*, 1986, doi: 10.1109/TMAG.1986.1064347.

[8] F. Preisach, "Über die magnetische Nachwirkung," *Zeitschrift für Phys.*, 1935, doi: 10.1007/BF01349418.

[9] M. Brokate, "Hysteresis operators," 1994.

[10] S. Bobbio, G. Miano, C. Serpico, and C. Visone, "Models of magnetic hysteresis based on play and stop hysterons," *IEEE Trans. Magn.*, 1997, doi: 10.1109/20.649875.

[11] A. A. Adly, "Accurate modeling of vector hysteresis using a superposition of preisach-type models," *IEEE Trans. Magn.*, 1997, doi: 10.1109/20.619694.

[12] G. Miano, C. Serpico, and C. Visone, "A new model of magnetic hysteresis, based on stop hysterons: an application to the magnetic field diffusion," *IEEE Trans. Magn.*, 1996, doi: 10.1109/20.497442.

[13] G. Friedman, "New formulation of the Stoner-Wohlfarth hysteresis model and the identification problem," *J. Appl. Phys.*, 1990, doi: 10.1063/1.344611.

[14] T. Matsuo, K. Ando, Y. Terada, and M. Shimasaki, "A Study of Representation of Hysteretic Characteristics by Stop and Play Models," *IEEJ Trans. Electron. Inf. Syst.*, 2003, doi: 10.1541/ieejeiss.123.1958.

[15] A. A. Adly and S. K. Abd-El-Hafiz, "Efficient modeling of vector hysteresis using a novel Hopfield neural network implementation of Stoner-Wohlfarth-like operators," *J. Adv. Res.*, 2013, doi: 10.1016/j.jare.2012.07.009.

[16] Paul Horowitz and H. Winfield, *The Art of Electronics*. Cambridge University Press, 1980.

[17] I. D. Mayergoyz, "Vector Preisach hysteresis models (invited)," *J. Appl. Phys.*, 1988, doi: 10.1063/1.340926.

[18] A. A. Adly and I. D. Mayergoyz, "A new vector Preisach-type model of hysteresis," *J. Appl. Phys.*, 1993, doi: 10.1063/1.353539.

[19] M. D'Aquino, C. Serpico, C. Visone, and A. A. Adly, "A New Vector Model of Magnetic Hysteresis Based on a Novel Class of Play Hysterons," *IEEE Transactions on Magnetics*, 2003, doi: 10.1109/TMAG.2003.816469.

[20] R. M. Bozorth, *Ferromagnetism*. 1993.

[21] V. K. Pecharsky and K. A. Gschneidner, "Magnetocaloric effect and magnetic refrigeration," *J. Magn. Magn. Mater.*, 1999, doi: 10.1016/S0304-8853(99)00397-2.

[22] K. A. Gschneidner and V. K. Pecharsky, "Magnetocaloric materials," *Annu. Rev. Mater. Sci.*, 2000, doi: 10.1146/annurev.matsci.30.1.387.

[23] V. K. Pecharsky, V. K. Pecharsky, K. A. Gschneidner, K. A. Gschneidner, A. O. Pecharsky, and A. M. Tishin, "Thermodynamics of the magnetocaloric effect," *Phys. Rev. B - Condens. Matter Mater. Phys.*, 2001, doi: 10.1103/PhysRevB.64.144406.

[24] N. A. de Oliveira and P. J. von Ranke, "Theoretical aspects of the magnetocaloric effect," *Physics Reports*. 2010, doi: 10.1016/j.physrep.2009.12.006.

[25] A. M. Tishin and Y. I. Spichkin, *The magnetocaloric effect and its applications*. 2016.

[26] V. Franco, J. S. Blázquez, J. J. Ipus, J. Y. Law, L. M. Moreno-Ramírez, and A. Conde, "Magnetocaloric effect: From materials research to refrigeration devices," *Progress in Materials Science*. 2018, doi: 10.1016/j.pmatsci.2017.10.005.

[27] M. Tejedor, J. A. Garcia, J. Carrizo, L. Elbaile, V. M. Prida, and J. D. Santos, "Positive and negative magnetostriction in nearly nonmagnetostrictive amorphous ribbons," *Japanese J. Appl. Physics, Part 1 Regul. Pap. Short Notes Rev. Pap.*, 2002, doi: 10.1143/jjap.41.4527.

[28] F. Vajda and E. Della Torre, "Vector moving hysteresis model with accommodation," *J. Magn. Magn. Mater.*, 1996, doi: 10.1016/0304-8853(95)00679-6.

[29] G. Friedman, Y. Huang, and J. Kouvel, "Toward experimental verification of vector Preisach hysteresis model," in *Digests of the Intermag Conference*, 2000, doi: 10.1109/intmag.2000.871799.

[30] D. C. Jiles and D. L. Atherton, "Theory of ferromagnetic hysteresis," *J. Magn. Magn. Mater.*, 1986, doi: 10.1016/0304-8853(86)90066-1.

[31] D. C. Jiles, J. B. Thoelke, and M. K. Devine, "Numerical Determination of Hysteresis Parameters for the Modeling of Magnetic Properties Using the Theory of Ferromagnetic Hysteresis," *IEEE Trans. Magn.*, 1992, doi: 10.1109/20.119813.

[32] K. Mehrotra, C. Mohan, and S. Ranka, *Elements of Artificial Neural Networks*. 2019.

[33] S. Haykin, "Neural networks: a comprehensive foundation by Simon Haykin," *The Knowledge Engineering Review*. 1999.

[34] A. A. Adly and S. K. Abd-El-Hafiz, "Utilizing neural networks in magnetic media modeling and field computation: A review," *Journal of Advanced Research*. 2014, doi: 10.1016/j.jare.2013.07.004.

[35] A. A. Adly and S. K. Abd-El-Hafiz, "Using neural networks in the identification of preisach-type hysteresis models," *IEEE Trans. Magn.*, 1998, doi: 10.1109/20.668057.

[36] C. Serpico and C. Visone, "Magnetic hysteresis modeling via feed-forward neural networks," *IEEE Trans. Magn.*, 1998, doi: 10.1109/20.668055.

[37] A. A. Adly, S. K. Abd-El-Hafiz, and I. D. Mayergoyz, "Identification of vector Preisach models from arbitrary measured data using neural networks," *J. Appl. Phys.*, 2000, doi: 10.1063/1.372853.

[38] C. Visone, C. Serpico, I. D. Mayergoyz, M. W. Huang, and A. A. Adly, "Neural-Preisach-type models and their application to the identification of magnetic hysteresis from noisy data," *Phys. B Condens. Matter*, 2000, doi: 10.1016/S0921-4526(99)00764-4.

[39] D. Makaveev, L. Dupré, M. De Wulf, and J. Melkebeek, "Isotropic vector hysteresis modeling with feed-forward neural networks," *J. Appl. Phys.*, 2002, doi: 10.1063/1.1456400.

[40] A. A. Adly, "Numerical Implementation and Testing of New Vector Isotropic Preisach-Type Models," *IEEE Trans. Magn.*, 1994, doi: 10.1109/20.334095.

[41] A. A. Adly and S. K. Abd-El-Hafiz, "Efficient implementation of vector preisach-type models using orthogonally coupled hysteresis operators," *IEEE Trans. Magn.*, 2006, doi: 10.1109/TMAG.2005.864095.

[42] A. A. Adly and S. K. Abd-El Hafiz, "Efficient implementation of anisotropic vector preisach-iype models using coupled step functions," in *IEEE Transactions on Magnetics*, 2007, doi: 10.1109/TMAG.2007.893702.

[43] J. V. Leite, N. Sadowski, P. Kuo-Peng, and J. P. A. Bastos, "A new anisotropic vector hysteresis model based on stop hysterons," in *IEEE Transactions on Magnetics*, 2005, vol. 41, no. 5, doi: 10.1109/TMAG.2005.845083.

[44] A. A. Adly and S. K. Abd-El-Hafiz, "Vector hysteresis modeling using octal clusters of coupled step functions," in *Journal of Applied Physics*, 2011, vol. 109, no. 7, doi: 10.1063/1.3563071.

[45] A. A. Adly and S. K. Abd-El-Hafiz, "Efficient vector hysteresis modeling using rotationally coupled step functions," in *Physica B: Condensed Matter*, 2012, vol. 407, no. 9, doi: 10.1016/j.physb.2011.05.053.

[46] A. A. Adly and S. K. Abd-El-Hafiz, "Speed-range-based optimization of nonlinear electromagnetic braking systems," *IEEE Trans. Magn.*, vol. 43, no. 6, pp. 2606–2608, 2007, doi: 10.1109/TMAG.2007.893411.

[47] J. Kennedy and R. Eberhart, "Proceedings of ICNN'95 - International Conference on Neural Networks," in *Particle Swarm Optimization*, 1995.

[48] M. Beleggia, M. De Graef, and Y. T. Millev, "The equivalent ellipsoid of a magnetized body," *J. Phys. D. Appl. Phys.*, vol. 39, no. 5, 2006, doi: 10.1088/0022-3727/39/5/001.

[49] V. F. Matyuk and A. A. Osipov, "Central demagnetization factor for bodies with different shapes. II. Rectangular rods," *Russ. J. Nondestruct. Test.*, vol. 36, no. 1, 2000, doi: 10.1007/BF02759390.

[50] A. A. Adly and S. K. Abd-El-Hafiz, "Vector hysteresis modeling in arbitrarily shaped objects using an energy minimization approach," *Appl. Comput. Electromagn. Soc. J.*, vol. 31, no. 7, 2016.

[51] A. A. Adly and S. K. Abd-El-Hafiz, "Utilizing four-node tetrahedra-shaped Hopfield neural network configurations in the local magnetization assessment of 3d objects exhibiting hysteresis," *AIP Adv.*, vol. 11, no. 2, 2021, doi: 10.1063/9.0000130.

[52] A. A. Adly and S. K. Abd-El-Hafiz, "An efficient hysteresis modeling methodology and its implementation in field computation applications," *J. Magn. Magn. Mater.*, vol. 434, 2017, doi: 10.1016/j.jmmm.2017.03.042.

[53] A. A. Adly and S. K. Abd-El-Hafiz, "Automated two-dimensional field computation in nonlinear magnetic media using Hopfield neural networks," in *IEEE Transactions on Magnetics*, 2002, vol. 38, no. 5 I, doi: 10.1109/TMAG.2002.803575.

[54] A. A. Adly and S. K. Abd-El-Hafiz, "Field Computation in Media Exhibiting Hysteresis Using Hopfield Neural Networks," *IEEE Trans. Magn.*, 2021, doi: 10.1109/TMAG.2021.3083424.

[55] L. Li, "High temperature magnetic properties of 49%Co-2%V-Fe alloy," *J. Appl. Phys.*, vol. 79, no. 8 PART 2A, 1996, doi: 10.1063/1.361732.

[56] A. A. Adly, "Simulation of field-temperature effects in magnetic media using anisotropic preisach models," *IEEE Trans. Magn.*, vol. 34, no. 4 PART 1, 1998, doi: 10.1109/20.706516.

[57] A. A. Adly, S. K. Abd-El-Hafiz, and I. D. Mayergoyz, "Using neural networks in the identification of Preisach-type magnetostriction and field-temperature models," *J. Appl. Phys.*, 1999, doi: 10.1063/1.369946.

[58] A. A. Adly and I. D. Mayergoyz, "Magnetostriction simulation using anisotropic vector Preisach-type models," *IEEE Trans. Magn.*, vol. 32, no. 5 PART 2, 1996, doi: 10.1109/20.539147.

[59] A. A. Adly, I. D. Mayergoyz, and A. Bergqvist, "Utilizing anisotropie preisach-type models in the accurate simulation of magnetostriction," *IEEE Trans. Magn.*, vol. 33, no. 5 PART 2, 1997, doi: 10.1109/20.619619.

[60] M. J. Sablik, D. C. Jiles, and L. Barghout, "First principles approach to magnetostrictive hysteresis (abstract)," *J. Appl. Phys.*, vol. 67, no. 9, 1990, doi: 10.1063/1.346066.

[61] M. J. Sablik and S. W. Rubin, "Relationship of magnetostrictive hysteresis to the ΔE effect," *J. Magn. Magn. Mater.*, vol. 104–107, no. PART 1, 1992, doi: 10.1016/0304-8853(92)90847-H.

[62] M. J. Sablik, "Relationship between magnetostriction and the magnetostrictive coupling coefficient for magnetostrictive generation of elastic waves," 2003, doi: 10.1063/1.1472986.

[63] J. B. Restorff, H. T. Savage, A. E. Clark, and M. Wun-Fogle, "Preisach modeling of hysteresis in Terfenol," *J. Appl. Phys.*, vol. 67, no. 9, 1990, doi: 10.1063/1.344708.

[64] A. A. Adly, I. D. Mayergoyz, and A. Bergqvist, "Preisach modeling of magnetostrictive hysteresis," *J. Appl. Phys.*, vol. 69, no. 8, 1991, doi: 10.1063/1.347873.

[65] A. Bergqvist and G. Engdahl, "A Stress-dependent Magnetic Preisach Hysteresis Model," *IEEE Trans. Magn.*, vol. 27, no. 6, 1991, doi: 10.1109/20.278950.

[66] A. A. Adly and S. K. Abd-El-Hafiz, "Identification and testing of an efficient hopfield neural network magnetostriction model," *J. Magn. Magn. Mater.*, vol. 263, no. 3, 2003, doi: 10.1016/S0304-8853(03)00066-0.

[67] A. A. Adly, D. Davino, and C. Visone, "Simulation of field effects on the mechanical hysteresis of Terfenol rods and magnetic shape memory materials using vector Preisach-type models," in *Physica B: Condensed Matter*, 2006, vol. 372, no. 1–2, doi: 10.1016/j.physb.2005.10.049.

[68] C. Visone, D. Davino, and A. A. Adly, "Vector preisach modeling of magnetic shape memory materials oriented to power harvesting applications," in *IEEE Transactions on Magnetics*, 2010, vol. 46, no. 6, doi: 10.1109/TMAG.2010.2042040.

[69] A. A. Adly and S. K. Abd-El-Hafiz, "Implementation of magnetostriction Preisach-type models using orthogonally coupled hysteresis operators," *Phys. B Condens. Matter*, 2008, doi: 10.1016/j.physb.2007.08.066.

[70] A. A. Adly, D. Davino, A. Giustiniani, and C. Visone, "Vector magnetic hysteresis modeling of stress annealed galfenol," *Journal of Applied Physics*, 2013, vol. 113, no. 17, doi: 10.1063/1.4798307.

[71] A. A. Adly and S. K. Abd-El-Hafiz, "Simulation of magneto-elastic materials using a novel vector hysteresis model," in *2016 13th International Conference on Electrical Engineering/Electronics, Computer, Telecommunications and Information Technology, ECTI-CON 2016*, 2016, doi: 10.1109/ECTICon.2016.7561269.

[72] A. A. Adly and S. K. Abd-El-Hafiz, "Construction of a magnetostrictive hysteresis operator using a tripod-like primitive hopfield neural network," *AIP Adv.*, vol. 8, no. 5, 2018, doi: 10.1063/1.5007009.

[73] A. A. Adly and S. K. Abd-El-Hafiz, "An Efficient Vector Hysteresis Model for Unidirectional Magneto-Elastic Interactions," *IEEE Trans. Magn.*, vol. 57, no. 2, 2021, doi: 10.1109/TMAG.2020.3024034.

[74] F. T. Calkins, R. C. Smith, and A. B. Flatau, "Energy-based hysteresis model for magnetostrictive transducers," *IEEE Trans. Magn.*, vol. 36, no. 2, 2000, doi: 10.1109/20.825804.

Index

Printed in the United States
by Baker & Taylor Publisher Services